# THE STATE
## *OF*
# FIRE

ALSO BY OBI KAUFMANN

*The Deserts of California: A California Field Atlas* (2023)

*The Coasts of California: A California Field Atlas* (2022)

*The Forests of California: A California Field Atlas* (2020)

*The State of Water: Understanding California's Most Precious Resource* (2019)

*The California Field Atlas* (2017)

# THE STATE OF FIRE

## Why California Burns

written and illustrated by
OBI KAUFMANN

Heyday, Berkeley, California

Library of Congress Cataloging-in-Publication Data

Names: Kaufmann, Obi, author.
Title: The state of fire : why California burns / written and illustrated by Obi Kaufmann.
Description: Berkeley, California : Heyday, [2024] | Includes bibliographical references.
Identifiers: LCCN 2024012011 (print) | LCCN 2024012012 (ebook) | ISBN 9781597146517 (hardcover) | ISBN 9781597146524 (epub)
Subjects: LCSH: Fire ecology--California. | Fire ecology--California--Maps. | Wildfires--California. | Wildfires--California--Prevention and control. | Forest fires--California. | Forest fires--California--Prevention and control.
Classification: LCC QH105.C2 K373 2024 (print) | LCC QH105.C2 (ebook) | DDC 577.2/409794--dc23/eng/20240416
LC record available at https://lccn.loc.gov/2024012011
LC ebook record available at https://lccn.loc.gov/2024012012

Cover Art: Obi Kaufmann
Cover Design: Ashley Ingram
Interior Design/Typesetting: Obi Kaufmann and Ashley Ingram

Published by Heyday
P.O. Box 9145, Berkeley, California 94709
(510) 549-3564
heydaybooks.com

Printed in East Peoria, Illinois, by Versa Press, Inc.

10 9 8 7 6 5 4 3 2 1

For the people of the future who will do the remembering

and for the people of now learning to be from here

and for the people of the past who should not be forgotten

## Ethical Presumptions of *The State of Fire*

There is no habitat space on Earth not touched by the industrialized hand of humanity, yet every human community is maintained by homeostatic parameters within an ecological architecture informed by ancient and fathomlessly complex biodiversity. The vectors of stress and resiliency that have historically maintained these parameters also represent thresholds of collapse should the architecture be utterly compromised.

Ecosystems are more than collections of living organisms. Everywhere, there are potentially infinite relationships that exist among conscious entities, and all of them require reciprocated effort to maintain the network of being.

the trees sing songs of fire older than
the rivers.

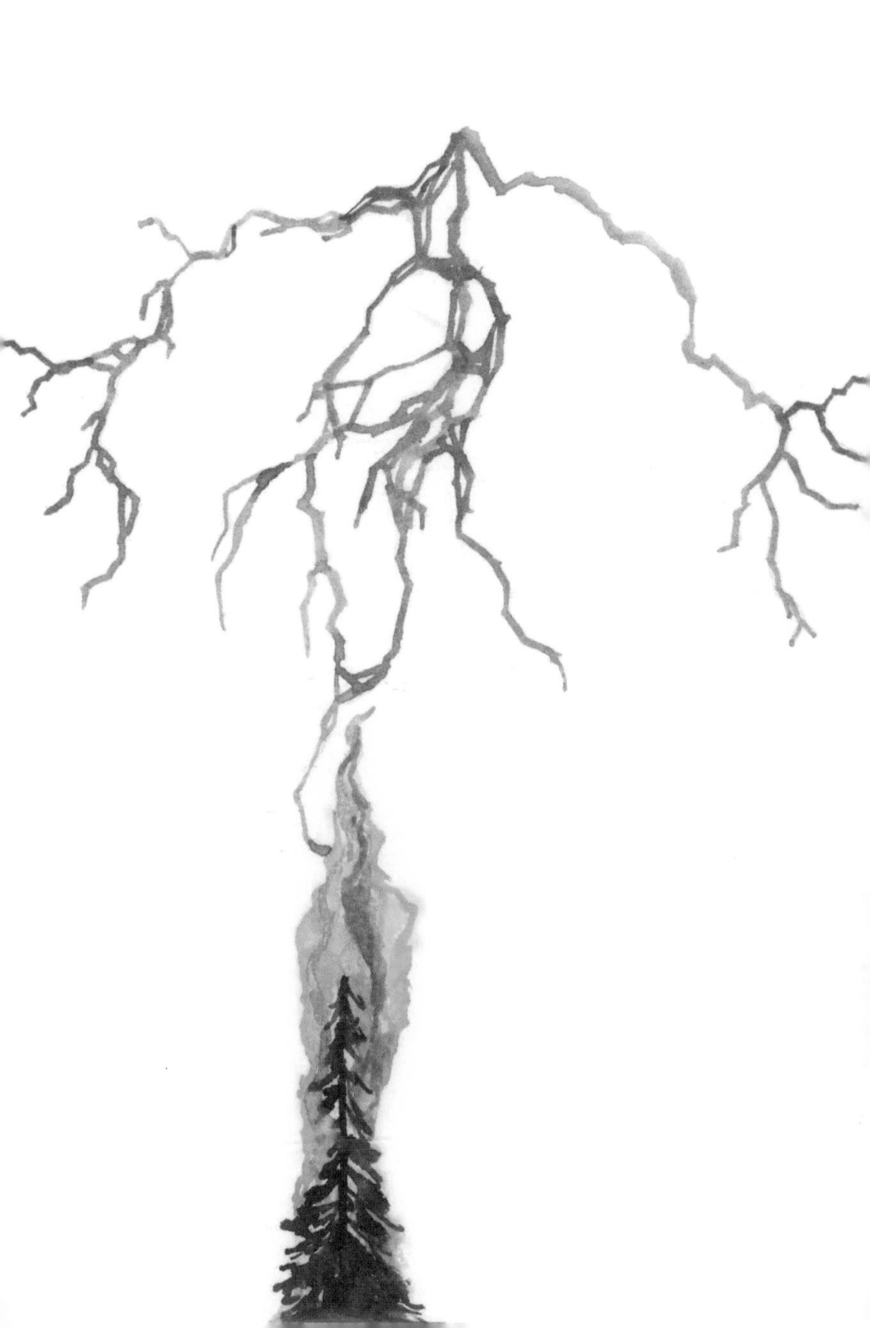

# CONTENTS

## Part Two: Fire Ecology

## Part Three: Fire Principles

exists in only one place in the park -
hidden inside a thick patch of other
grasses - north of 299, Tower house
historic district.

05.17.23

field sketch
by obykauffmann

Shasta Gilia

Gilia sp.
(species name
pending)

bractlet

flower 4mm
diameter

whiskeytown
NRA

Polemoniaceae - 400 species
worldwide
25 genera

# INTRODUCTION

## Searching for Nature in the Anthropocene

### *Spring in Whiskeytown*

It took me a few minutes to figure out what I was looking at, and significantly longer to figure out how I felt about it. Standing next to the reservoir on a sunny day, surrounded by human recreationalists, I was not in any so-called wilderness. I was at the edge of the *wildland–urban interface*, a term used until very recently only by policy wonks and land managers, but now relevant and popular, referring to where the built world of infrastructure meets untended landscapes in which few humans currently live. Recently, fire had left this land black and barren, and I'm sure it did the same to so many homes of the people who now relaxed on the beach. That day, both the land and the people seemed to have let the trauma subside and embraced the simple joy of cool, spring sunshine. A dense grove of green, broadleaved trees bent from the shore, offering shade to the beachgoers, and a snow-melt creek babbled nearby, flanked by purple and white flower-crowned shrubs attended by a host of happy birds and insects. In the distance, vernal snow still clung to Shasta Bolly's six-thousand-foot cornice, which wore a skirt of brown conifers, evidence of an ecosystem still deliberating the character of its postfire reorganization.

But even as I pondered processes that once may have been beyond the reach of human influence, I couldn't shake the historical question:

Was any of it *not* shaped by human artifice? Every species in this watershed existed in some several-thousand-year relationship to human design and effort. And post–gold rush American will had manipulated, regulated, impounded, deformed, and simplified the patterns of the living world here. At the heart of Whiskeytown National Recreation Area (WNRA), five years after megafire (the Carr Fire) reset and transformed this ecosystem, I found a landscape emblematic of what was happening across the planet. On that beautiful day, the question chilled me: Now that every terrestrial ecosystem was subject to the industrialized legacy of humanity, wasn't the whole world now inside the wildland–urban interface?

That legacy transforms the local landscape through the air, as pollution and climate breakdown by way of global warming; across the earth, through fragmentation by roadways and infrastructure buildup at the wildland–urban interface—an ever-increasing area of fire-prone rural and natural landscapes encroached on by sprawling modern settlement patterns; in the water, through the impounding of riverine ecosystems and dramatically altering hydrology. And then there is fire. The past, present, and future of fire are key in understanding humanity's power to upend the world. Ethical questions about what nature is pile up: What parameters of wildness define nature? What are our responsibilities to maintain those parameters? To what degree does fire define California's natural world? Are big fires, like the Carr Fire, indicative of the natural world restoring itself? Are we inside an inevitable age of reckoning? Or are these fires symptomatic of ill health and imbalance representing a landscape-scale dysbiosis, or a kind of disease that leads to little but ruinous calamity? Does the land experience trauma as we do? If we put effort into understanding how, where, and why California burns, might that be a prescription of healing for both the land and for ourselves?

In the spring of 2023, having been accepted into the artist residency program of the National Park Service (NPS) at WNRA, I knew I was going to one of California's best field laboratories for empirically researching these questions. Park staff welcomed me on a months-long cook's tour of this unique parcel of land. I joined students and scientists, rangers and researchers, and entered a season of learning that would change my life. WNRA is sixtyish square miles in size, ten miles west of Redding, California. Between the Sacramento River Valley to the east, Shasta to the north, the Trinity Mountains to the west, and the coastal ranges to the south, WNRA has unique geologic and climatic qualities found nowhere else in the state. The reservoir on Clear Creek called Whiskeytown Lake is flanked by four-hundred-million-year-old soils that get blasted with sixty inches of rain in the winter, then cooked under weeks of temperatures that soar to more than 110 degrees in the summer. Before I arrived, I knew that I would be enmeshed in an emerging landscape because I knew what had happened five years prior to my visit. I knew all about the Carr Fire.

It was the last week of July in 2018 when the Carr Fire bore down on Whiskeytown and began its month-long crawl over a quarter of a million acres, an area eight times the size of WNRA. Before 2023, what I knew of the Carr Fire came from what I heard about on the news, which focused—and only ever seems to focus—on the human cost: damage and lives lost. Three days after it initially ignited, the Carr Fire spawned an apocalyptic fire tornado in a suburban neighborhood of Redding. At 2,700° F and blowing at 165 miles per hour, the tornado existed for thirty minutes, was a thousand feet across and three miles high, and obliterated a half-mile-long stretch of homes. The damage was equivalent to a Category 5 hurricane and exerted the localized force of several atomic weapons equivalent to those dropped on Hiroshima and

Nagasaki in 1945. Five people died. Eventually, the Carr Fire would take eight lives in total. Across the surrounding woodlands, the severe burn zone, or the area of 100 percent plant mortality, extended for tens of thousands of acres. But the Carr Fire was not the largest or most severe fire in California that year, the year before, or the year after. In the past twenty years, the Carr Fire doesn't even make the top ten.

In the wake of such terror, it is easy to imagine that twenty-first-century wildfire is symbolic of some grand failure. Or to assume that California's evergreen development has reached a limit, and that society's relationship with the more-than-human world needs a grand reset. Both ideas may indeed be true. Whatever the case may be, my tour of WNRA made it startling clear that California's ecology and biodiversity have an ancient, profound, and fecund relationship with human fire on the land. The relationship between anthropogenic fire and the character of California's living landscape is so intrinsic that without it, California would not be California. Fire is a primary component of what is referred to as traditional ecological knowledge, which is a bundle of regionally specific, indigenous technologies of ecological fecundity on the land. Without millennia after millennia of applied indigenous fire, and without traditional ecological knowledge, California would not demonstrate the world-class biodiversity that it still does today. California's biodiversity largely exists as it does because of fire and not despite it; that fact is the thematic core of this book.
The human story of Whiskeytown's landscape is California's story, and is a very old story of resiliency, community, and adaptability. Despite centuries of pollution and injury, evidence of extant, robust biodiversity is everywhere. Five years after the megafire, botanists are combing the hills, redrawing plant distribution maps and discovering new species of fire-following flower species. The resurgent, vernal woodland sets a poetic

backdrop for a grand city of songbirds and drunken bumblebees madly attending their wildflower paradise. The surprisingly vibrant attitude of the snag forest, the postfire forest ecosystem, is sometimes disguised by the gray-and-black color of its bone-like tree trunks now claimed as avian apartment buildings. Inside, untold thousands of nests hold a million dreaming infant birds, awaiting their turn to sing in the morning sun. Across the baked, greenstone soil of the Whiskeytown chaparral is a delightful panoply of life, an ecosystem catching its breath from the gut-punch fire. The ecosystem not only understands that fire will return but wishes for fire's recurrence, in accordance with how things once were and shall be again. Even though, with the advent of so many novel stressors, from invasive grass to climate breakdown, fire's regular return has been forever altered from its precontact regime, evidence abounds for how this place is remembering itself, and fire is the agent of that reset, at once ancient and new.

Obi Kaufmann
Oakland, California
Spring 2024

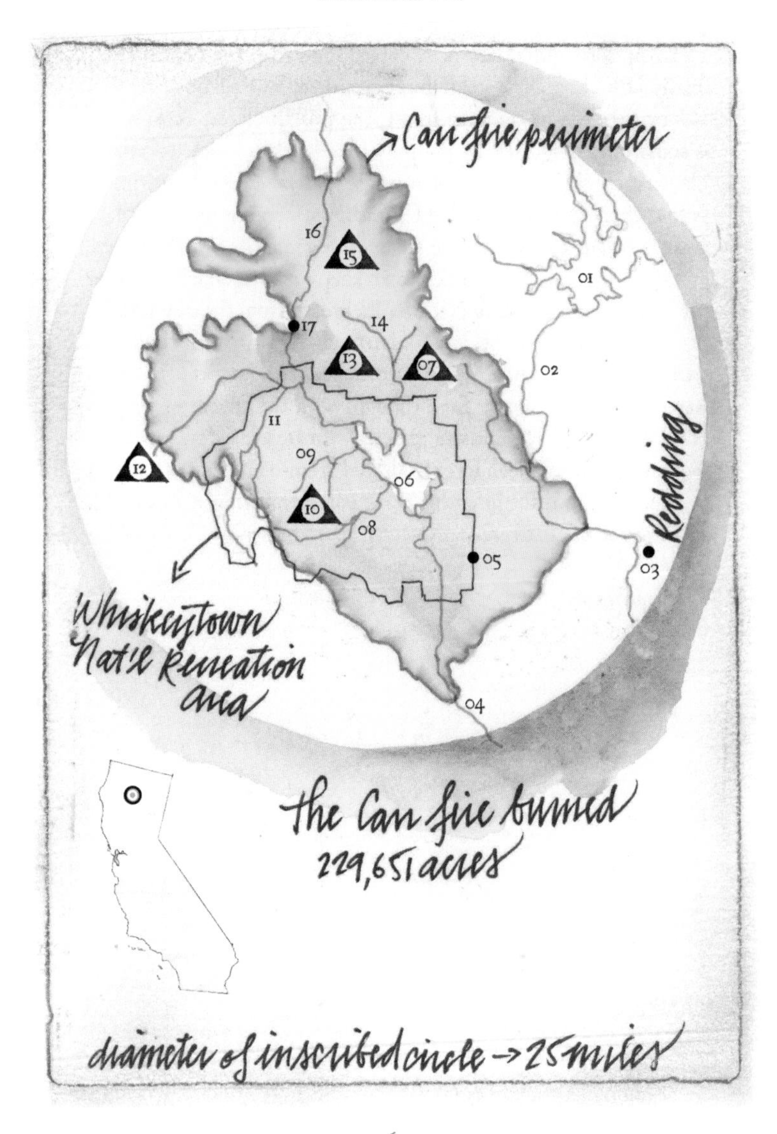
→ Carr fire perimeter
16
15
01
17
14
13
07
02
11
12
09
06
10
08
Redding
05
03
04
Whiskeytown
Nat'l Recreation
area
the Carr fire burned
229,651 acres
diameter of inscribed circle → 25 miles

# Map 00.01 Carr Fire and Whiskeytown National Recreation Area

01. Shasta Lake
02. Sacramento River
03. Redding
04. Clear Creek
05. Border of the WNRA
06. Whiskeytown Lake
07. South Fork Mountain Lookout 3,447'
08. Brandy Creek
09. Boulder Creek
10. Shasta Bolly 6,160'
11. Crystal Creek
12. Buckhorn Summit 3,913'
13. Mad Mule Mountain 3,066'
14. Whiskey Creek
15. Shirttail Peak 4,100'
16. Clear Creek
17. French Gulch

## Note about Plant Families

The world is held together by relationships between living beings. Energy within any and every ecosystem is passed around in cycles that move in all directions through food webs and across habitat spaces. Biodiversity is the measure of the amount of species and is also related to the resiliency of any ecosystem's ability to withstand disturbance or damage. Because plants can produce their own nutrient energy, they form the basis of all ecological models, and the more biodiversity in plant types, the stronger the ecological framework. What follows is a complete list of the plant families—groups of taxa (plant types arranged by taxonomic families that include hundreds of individual species) present in the valley and what are known as the inland coast ranges, or the ecological geography of where WNRA is located. This list is a roll call, or a roster of attendance for all the valuable members of this more-than-human community. Every one of these families is priceless. Every one of these families is important. Every one of these families has an ancient relationship with fire. Whether that fire is measured by its intensity (how hot it burns), its severity (how much damage it does), or its return (how often it comes back), fire dictates the local population, the configuration, and the distribution of all plant communities—and therefore all animal communities. What if a single plant family is missing? Can the consequences be measured? How many pieces of ecological architecture can go missing before the whole system collapses? Any trend toward simplification and away from biodiversity will trigger a cascading set of potentially disastrous consequences and is one of the many reasons why a greater understanding of fire impact in the age of global warming is so important to address.

(Native) (invasive) (naturalized) Page 001

Plant families of Whiskeytown with common example species

| | | |
|---|---|---|
| Aceraceae | Maple family | Acer macrophyllum<br>big leaf maple |
| Alliaceae | onion family | Allium membranaceum<br>papery onion |
| Anacardiaceae | sumac family | Toxicodendron diversilobum<br>poison oak |
| Apiaceae | carrot family | Foeniculum vulgare<br>fennel |
| Apocynaceae | dogbane family | Apocynum fascicularis<br>whorled milkweed |
| Asteraceae | sunflower family | |
| | asterae - aster tribe | Eriophyllum lanatum<br>wooly sunflower |
| | cynarae - thistle tribe | Cirsium cymosum<br>peregrine thistle |
| | Cichoriaea - chickory tribe | Tragopogon dubius<br>yellow salsify |
| | sunflower tribe | Wyethia helenioides<br>wooley mule's ear |
| | sneezeweed subtribe | Eriophyllum confertiflorum<br>western yarrow |
| | everlasting tribe | Anaphalis margaritacea<br>pearly everlasting |
| Berberidaceae | Barberry family | Berberis nervosa<br>oregon grape |
| Boraginaceae | Borage family | Amsinckia menziesii<br>common fiddleneck |

page 002

| | | |
|---|---|---|
| Brassicaceae | mustard family | Streptanthus tortuosus<br>common jewel flower |
| Campanulaceae | bell flower family | Githopsis specularioides<br>common bluecup |
| Caprifoliaceae | honeysuckle family | Sambucus nigra<br>blue elderberry |
| Caryophyllaceae | Pink family | Silene californica<br>indian pink |
| Convolvulaceae | morning glory family | Calystegia occidentalis<br>western morning glory |
| Cornaceae | dogwood family | Cornus nuttallii<br>Pacific dogwood |
| Crassulaceae | stonecrop family | Sedum spathulifolium<br>common stonecrop |
| Cucurbitaceae | Gourd family | Marah fabacea<br>California man-root |
| Cupressaceae | cypress family | Calocedrus decurrens<br>incense cedar |
| Cyperaceae | sedge family | Carex amplifolia<br>big leaf sedge |
| Ericaceae | heather family | Arctostaphylos manzanita<br>common manzanita |
| Fabaceae | pea family | Lupinus nanus<br>sky lupine |
| Fagaceae | oak family | Notholithocarpus densiflorus<br>tanoak |
| Geraniaceae | geranium family | Erodium botrys<br>broadleaf filaree |

Page 003

| | | |
|---|---|---|
| Grossulariaceae | gooseberry family | Ribes malvaceum<br>chaparral currant |
| Hippocastanaceae | horse chestnut family | Aesculus californica<br>California buckeye |
| Hydrophyllaceae | waterleaf family | Eriodictyon californicum<br>yerba santa |
| Juncaceae | rush family | Juncus exiguus<br>Klamath rush |
| Lamiaceae | mint family | Salvia sonomensis<br>sonoma sage |
| Lauraceae | laurel family | Umbellularia californica<br>California bay |
| Liliaceae | lily family | Calochortus elegans<br>elegant star tulip |
| Loasaceae | blazing star family | Mentzelia laevicaulis<br>common blazing star |
| Malvaceae | mallow family | Fremontodendron californicum<br>flannel bush |
| Oleaceae | olive family | Fraxinus latifolia<br>oregon ash |
| Onagraceae | evening primrose family | Clarkia rhomboidea<br>farewell-to-spring |
| Papaveraceae | poppy family | Eschscholzia californica<br>California poppy |
| Pinaceae | pine family | Pinus attenuata<br>knobcone pine |

Page 004

| | | |
|---|---|---|
| Poaceae | grass family | |
| | Agrostideae bent-grass tribe | |
| | Stipa needlegrass | |
| | avena oats | |
| | Aveneae oat tribe | |
| | Festuceae fescue tribe | |
| | Melica melics | |
| | Bromus bromes | |
| | Hordeae | |
| | Hordeum foxtail | |
| Polemoniaceae | phlox family | phlox speciosa showy phlox |
| Polygonaceae | buckwheat family | Eriogonum umbellatum sulphur buckwheat |
| Portulacaceae | purslane family | Claytonia perfoliata miner's lettuce |
| Primulaceae | primrose family | Dodecatheon hendersonii Henderson's shooting stars |
| Ranunculaceae | buttercup family | Aquilegia formosa red columbine |
| Rhamnaceae | buckthorn family | Ceanothus integerrimus deerbrush |
| Rosaceae | rose family | Cercocarpus betuloides mountain mahogany |
| Rubiaceae | madder family | Cephalanthus occidentalis button-willow |

"If we don't remember their names, they are not going to remember ours and we will all forget how they feed and take care of the people." Greg Sarris
Southern Pomo

The evolutionary history is as old as terrestrial life and endemic to our planet's biosphere, and we, the humans, are the species of fire. The thing that burning hydrocarbons is and does means and has meant everything to the functioning of every human society ever. Fire is more than a kind of symbiote, it is an emergent function of humanity's relationship to the world. Getting right with fire on the land will determine the quality of humanity's residency in California for centuries to come.

Streptanthus
Jewel Flower

obikaufmann
nps_whis
May 23

- best acorns
- can live 500 years

California black oak
Quercus kelloggii
(red oak section
Quercus sect. Lobatae)
deciduous

at the beginning of may, snow drop bush was everywhere, a grand cascade of white flowers - by the end of may, all the blooms on this evergreen shrub were gone. May 23

"the words end, but the meaning continues." - Basho

California snow drop bush
Styrax redivivus

obkaufmann

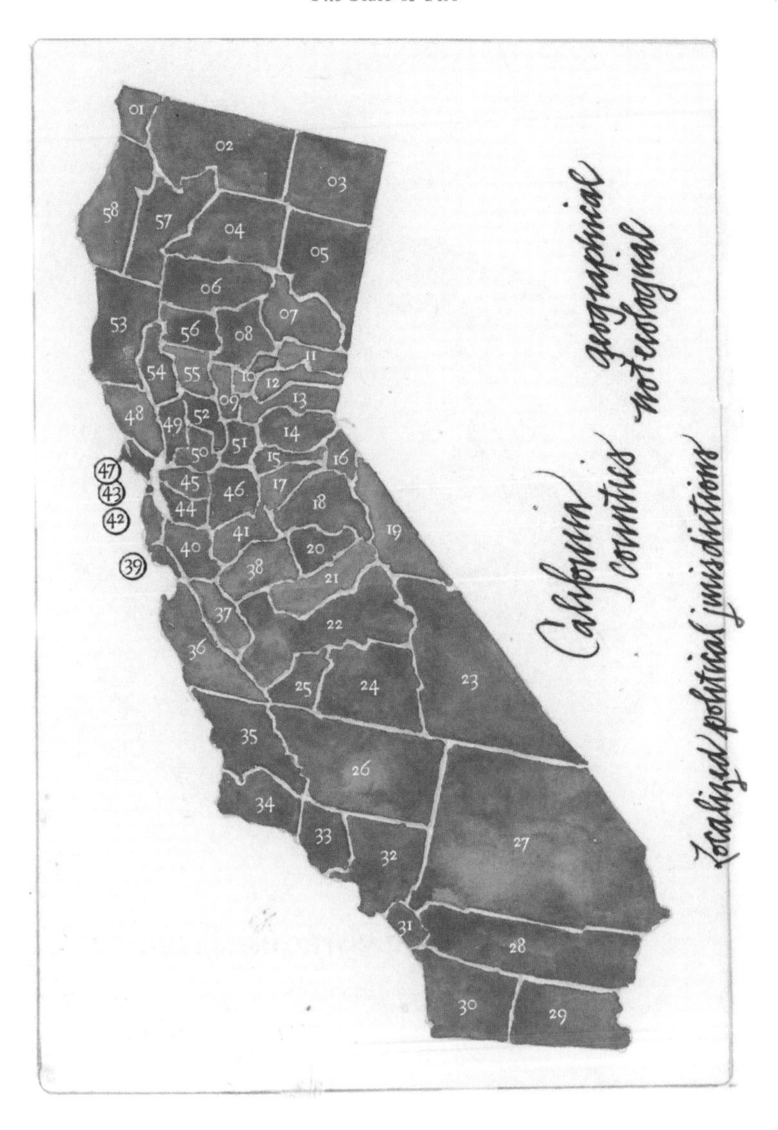
California
counties
geographical
not ecological
localized political jurisdictions
01
02
03
04
05
06
07
08
09
10
11
12
13
14
15
16
17
18
19
20
21
22
23
24
25
26
27
28
29
30
31
32
33
34
35
36
37
38
39
40
41
42
43
44
45
46
47
48
49
50
51
52
53
54
55
56
57
58

## Map 00.02 California Counties

01. Del Norte
02. Siskiyou
03. Modoc
04. Shasta
05. Lassen
06. Tehama
07. Plumas
08. Butte
09. Sutter
10. Yuba
11. Sierra
12. Nevada
13. Placer
14. El Dorado
15. Amador
16. Alpine
17. Calaveras
18. Tuolumne
19. Mono
20. Mariposa
21. Madera
22. Fresno
23. Inyo
24. Tulare
25. Kings
26. Kern
27. San Bernardino
28. Riverside
29. Imperial
30. San Diego
31. Orange
32. Los Angeles
33. Ventura
34. Santa Barbara
35. San Luis Obispo
36. Monterey
37. San Benito
38. Merced
39. Santa Cruz
40. Santa Clara
41. Stanislaus
42. San Mateo
43. San Francisco
44. Alameda
45. Contra Costa
46. San Joaquin
47. Marin
48. Sonoma
49. Napa
50. Solano
51. Sacramento
52. Yolo
53. Mendocino
54. Lake
55. Colusa
56. Glenn
57. Trinity
58. Humboldt

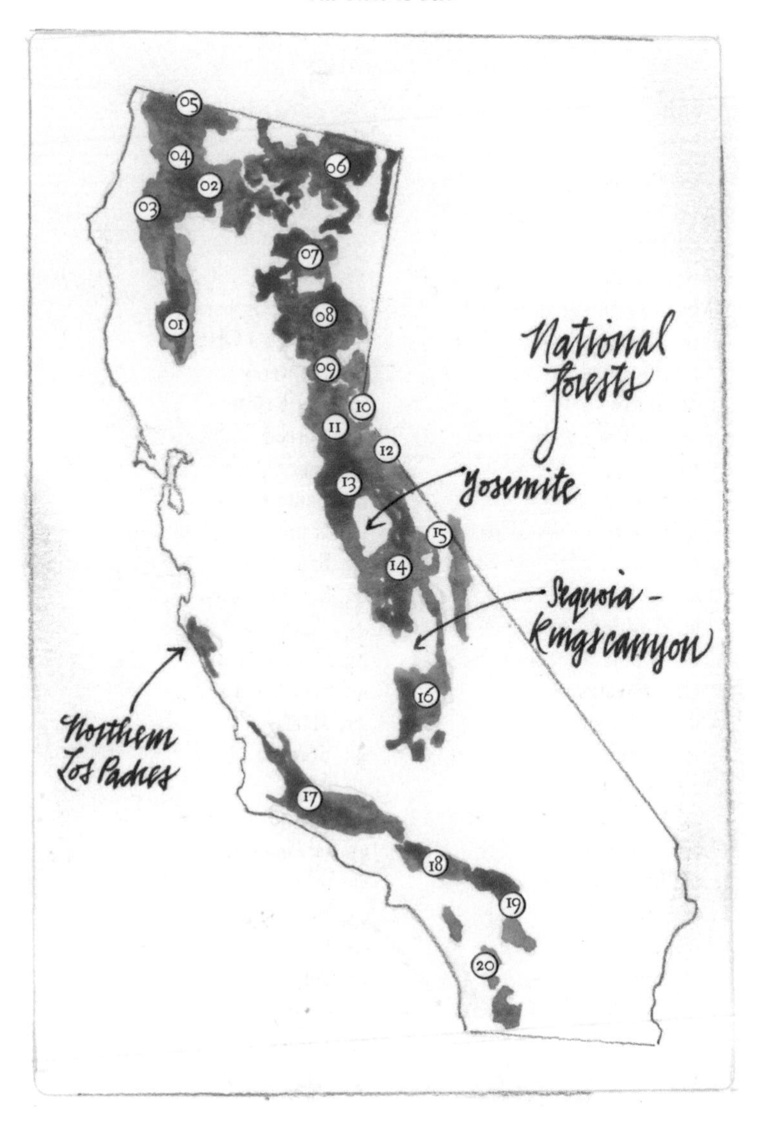
National Forests
Yosemite
Sequoia-Kings Canyon
Northern Los Padres
01
02
03
04
05
06
07
08
09
10
11
12
13
14
15
16
17
18
19
20

## Map 00.03 National Forests in California

Federal, public land managed by the United States Forest Service for energy, livestock, recreation, and timber—"Land of Many Uses." Excludes lands of the National Park Service and the Bureau of Land Management.

01. Mendocino National Forest
02. Shasta Trinity National Forest
03. Six Rivers National Forest
04. Klamath National Forest
05. Rogue River National Forest
06. Modoc National Forest
07. Lassen National Forest
08. Plumas National Forest
09. Tahoe National Forest
10. Lake Tahoe Basin Management Unit
11. El Dorado National Forest
12. Humboldt Toiyabe National Forest
13. Stanislaus National Forest
14. Sierra National Forest
15. Inyo National Forest
16. Sequoia National Forest
17. Los Padres National Forest
18. Angeles National Forest
19. San Bernardino National Forest
20. Cleveland National Forest

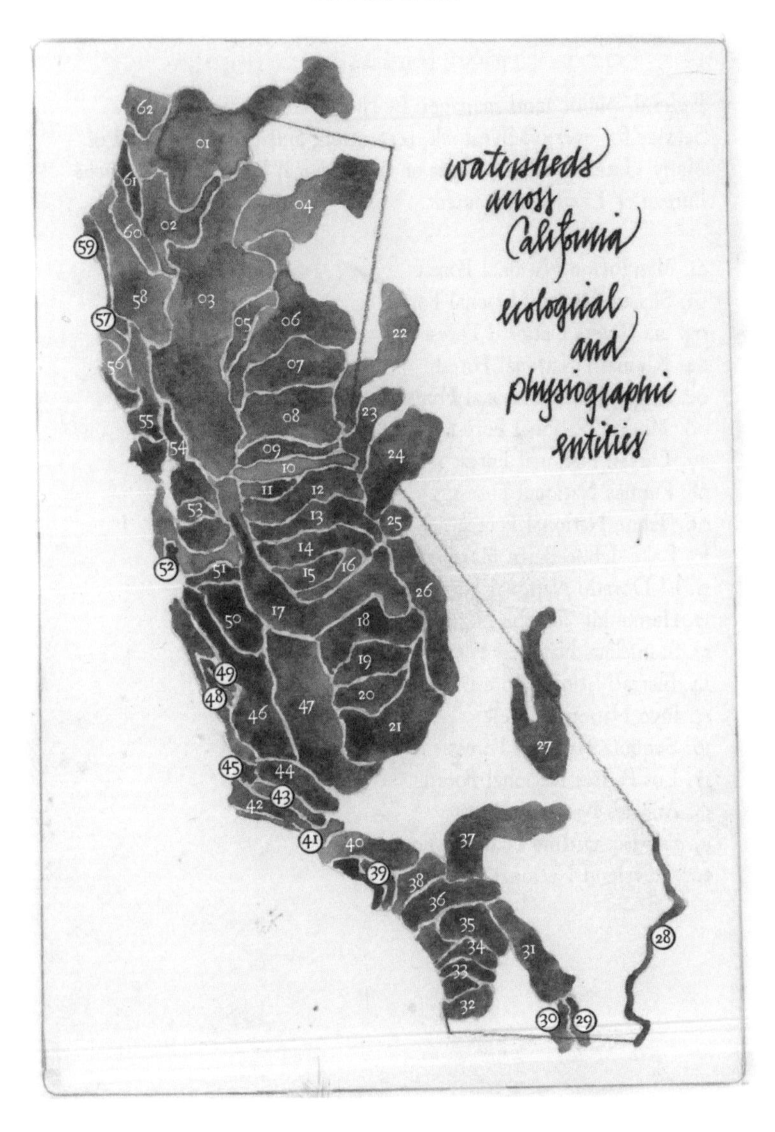
watersheds
across
California
ecological
and
physiographic
entities
01
02
03
04
05
06
07
08
09
10
11
12
13
14
15
16
17
18
19
20
21
22
23
24
25
26
27
28
29
30
31
32
33
34
35
36
37
38
39
40
41
42
43
44
45
46
47
48
49
50
51
52
53
54
55
56
57
58
59
60
61
62

## Map 00.04 California's Major Watersheds

Regional drainage basins, bordered by ridges or other landscape features—indicative of integral ecological physiography and of generalized connective biogeography.

01. Klamath River
02. Trinity River
03. Sacramento River
04. Pit River
05. Butte Creek
06. Feather River
07. Yuba River
08. American River
09. Cosumnes River
10. Mokelumne River
11. Calaveras River
12. Stanislaus River
13. Tuolumne River
14. Merced River
15. Chowchilla River
16. Fresno River
17. San Joaquin River
18. Kings River
19. Kaweah River
20. Tule River
21. Kern River
22. Truckee River
23. Carson River
24. Walker River
25. Mono Lake
26. Owens River
27. Amargosa River
28. Colorado River
29. Alamo River
30. New River
31. Whitewater River
32. San Diego River
33. San Luis Rey River
34. Santa Margarita River
35. San Jacinto River
36. Santa Ana River
37. Mojave River
38. San Gabriel River
39. Los Angeles River
40. Santa Clara River
41. Ventura River
42. Santa Ynez River

43. Sisquoc River
44. Cuyama River
45. Santa Maria River
46. Salinas River
47. Tulare Lake
48. Nacimiento River
49. San Antonio River
50. San Benito River
51. Pajaro River
52. San Lorenzo River
53. Alameda Creek
54. Napa River
55. Russian River
56. Navarro River
57. Noyo River
58. Eel River
59. Mattole River
60. Mad River
61. Redwood Creek
62. Smith River

watching across the cinders
millennia that burn and crumble
Where wildflowers take down mountains
energetic power flows from chemistry to biology
and back within the geosynchronous
pendulum
of life's blinking eye

# PART ONE: FIRE HISTORY

## 01.01 Endemic/Indigenous Pyrosymbiosis

*The evolution of fire in California*

Because the fuel is contingent on life-specific criteria—plant-based hydrocarbons—fire may be unique to this planet, or at least as rare as life is across the cosmos. Nearly three billion years ago, the evolutionary invention of photosynthesis in microbial cyanobacteria set the stage for fire's emergence. By transforming the atmosphere from a methane-rich environment suitable only for supporting anaerobic life to an oxygen-rich environment, aerobic metabolisms inside what would emerge as multicellular organisms were able to form, fueling an evolutionary race toward a complexity of life forms that continues to unfold today.[1] During the transformation of the atmosphere, called the Great Oxidation Event, mycelial hyphae (fungal tissue) began a billion-year march of colonizing inorganic soils on dry land, preparing the ground for the coming of the first land plants, green algae, six hundred million years ago.[2] The first hydrocarbons that fed the first lightning-sparked land fires were probably mycelial, although the earliest fossilized charcoal (called fusain) that has ever been found and is generally regarded as the first evidence of fire corresponds with the first established plant-based ecosystems from the early Devonian period, a bit more than four hundred million years ago.[3]

The planetary forces that established deep-time fire regimes were already ancient when California began its long trek to resembling its current geography. Two hundred million years ago, the formation of the granite batholith that would eventually rise as the Sierra Nevada began its process of formation. One hundred million years ago, as the now subsumed Farallon tectonic plate was ground under the North American tectonic plate, the mountains first rose as a chain of volcanoes that ecologically isolated what lands would emerge to the west of the range from the rest of North America.[4] Just about the time that grass (Poaceae family) made its evolutionary debut, before the Cretaceous Tertiary extinction event responsible for the death of all nonavian dinosaurs, the Sierra Nevada topped out at less than eight thousand feet and was covered with arboreal populations that would eventually become today's pines, cedars, and firs.[5] Without grass, and tens of millions of years before the red oak clade of what eventually became today's oaks (family Quercus) migrated into the proto–California Floristic Province, wildfire would not have had the kind of fuel provided by the understory of arboreal habitats today.[6] Because of this and because there is little consensus regarding global oxygen levels during the Paleogene period (66–23 million years before present—MYBP), long before California was the California of today, any conclusions about fire patterns in this time frame are, for now, purely speculative.[7]

California was a vastly different place before the coming of humanity. Throughout the Cenozoic (from 66 MYBP until now), the so-called age of mammals, California slowly began to resemble its current self. In the late Miocene and early Pliocene (about 5 MYBP to 3 MYBP), after the coastal ranges had neared their current height, and after the San Andreas Fault had long established its contemporary configuration as a transform fault, the California Current established itself in the eastern Pacific and begin to deliver to California what would eventually be a key component of today's Mediterranean climate pattern. But there was an interruption: a long period of freeze and thaw across a couple of million years that would not break until a time that coincided with humanity's arrival in California. Some paleobiologists have described the ice age of the Pleistocene as a time of the California Serengeti. A whole parade of huge predators, from saber-toothed catlike creatures (*Smilodon* sp., twice as large as any modern cat in the family Felidae) to the largest bear to ever stalk the Americas (*Artodus simis*, four times the size of California's modern black bear), would prey on over twenty herbivores that weighed more than five hundred pounds. Mammals such as camels, mammoth, sloth, bison, horses, and tapirs subjected the grassy woodlands and old-growth forests of Pleistocene California to a grazing regime that would have dictated fuel supplies across California. The grazing pressure would have selected for adaptive plants that quickly responded to injury with vigorous growth, a trait that would be beneficial after the late-Pleistocene megafaunal extinctions and the coming of human fire practices. Grazing pressure may have been what established the future adaptations of pyrosymbiosis, propagated by tribal peoples in the millennia to come. On the other side of the globe, it was *Homo erectus*, about one and a half million years ago, that was probably the first human species to learn the secret of the campfire—one million years before *Homo sapiens* made its appearance.

The ice age of the Pleistocene did not release its long grip on the Sierra Nevada until the subsidence of the Last Glacial Maximum, into the current era of interglaciation, called the Holocene. The earliest archaeological evidence for humanity's arrival in California is at the Calico Hills site in the Mojave Desert.[8] Tools at the site are generally agreed to be from a culture that inhabited this late-Pleistocene site about 20,000 years BP (before present). In the time window between 20,000 and 12,000 BP, many large animal species—such as the grizzly bear, moose, wolf, and elk—immigrated to North America across the land bridge from Asia. All found a home in California and developed as divergent, endemic species. The last California moose was killed about seven hundred years ago; the grizzly and the wolf held on until the twentieth century (although the wolf may be making a comeback), and the California elk, called the tule elk, who grazed these oak woodlands, remains extant in coastal reserves and is experiencing a population resurgence because of conservation efforts.

California—by way of its positioning on the globe (the west coast of a major land continent, north of the tropics, and subject to the California Current), its physiography (sequestered from the rest of the North America botanical provinces by the wall that is the Sierra Nevada), and its climate (one of six Mediterranean climate locales in the world)—has been sculpted by fire in unique ways for the last several million years. After the coming of human populations, pyrogenic economies began to establish themselves, culminating in a stabilizing epoch that began about five thousand years ago. From then until the Euro-American colonization, California's forest habitats remained consistent primarily because of Indigenous fire application. This is why every historically native botanical community in California has a relationship to fire's regular return. And it's why California would not be California, at least as we recognize it, without fire. Fire means everything to California. Since the ice age it has been an ever-present character in the theater of evolution. Fire has sculpted every arboreal, shrubby, and grassland habitat within the California Floristic Province. It has informed coevolved ecological adaptation strategies for tens of thousands of years. And the human hand has been there for that grand span of time since the ice age, guiding that coevolution to coax the land to regularly deliver the nutrient return and the ecological services that fire disturbance provides, helping to sustain what was to be the largest population of North American peoples at the time of contact. Over the centuries following colonization, the attempted genocide of California's Indigenous peoples, the suppression of traditional ecological knowledge, the ubiquity of invasive grass species, and the extractive commodification of California's natural resources have led to widespread, chaotic, and destructive wildfire that has no analog throughout the entire climate history of the Holocene epoch.

In the first two decades of the twenty-first century, California endured the largest, most severe, and destructive period of statewide conflagration events recorded up to that point. Now two stories are emerging that shed light on the meaning of fire in California. The first is a very old one, embodied today in resurgent Indigenous technologies that use fire to restore and reciprocate with the land's potentially infinite capacity for renewal. The second is a very new story, one that tells how the study of fire ecology is changing in academic and policy arenas and distinguishes this discipline from the timber-oriented realm of forestry and agriculture.

under the self-pruning
Ponderosa pine
made for this...
Top of the understory
forest floor
walking California as world maker
anthropogenic landscapes (the nonwild)
the false-wild
fire : hearth : <domos>
domesticated
fire = home
no "wildland"
evidence of emergent landscapes
human/fire-made
symptoms
- no large predators
- over-crowded fire-prone vegetation
- fragmenting roads
- livestock
- litter everywhere
- giardia
- infrastructure : cell service
- pathogens
- invasive plants
- altered climate
- crows, ravens, raccoons, coyote
agents of Anthropos
"what you tame, you become responsible for, forever" - the Little Prince

## *Three fires in the age of loneliness*

The long history of fire, according to fire historian Stephen Pyne, can be divided into three epochs.[9] *First fire* is original fire on the land—lightning- and volcano-sparked fire that knew no guiding hand other than the sources of its fuel and the oxygen ratio in the air. *Second fire* is that of the Native people who employ landscape fire and hearth fire for community profit. *Third fire* is recent and represents the burning of hydrocarbons mined from the earth in the form of fossil fuels. Third fire is what the global culture of humanity relies on today, and it might just be the force that, in the end, consumes its creator. Humanity's relationship with fire is integral, such that it has been suggested by fire journalist John Vaillant that the species should call itself *Homo flagrens* and that the current age should not be termed the Anthropocene but the Pyrocene or, more specifically to third fire, the Petrocene.[10] It has also been suggested by the naturalist E. O. Wilson that given the dramatic decline in biodiversity around the world, partly because of third fire, the present age would most appropriately be called the Eremocene, the age of loneliness.[11]

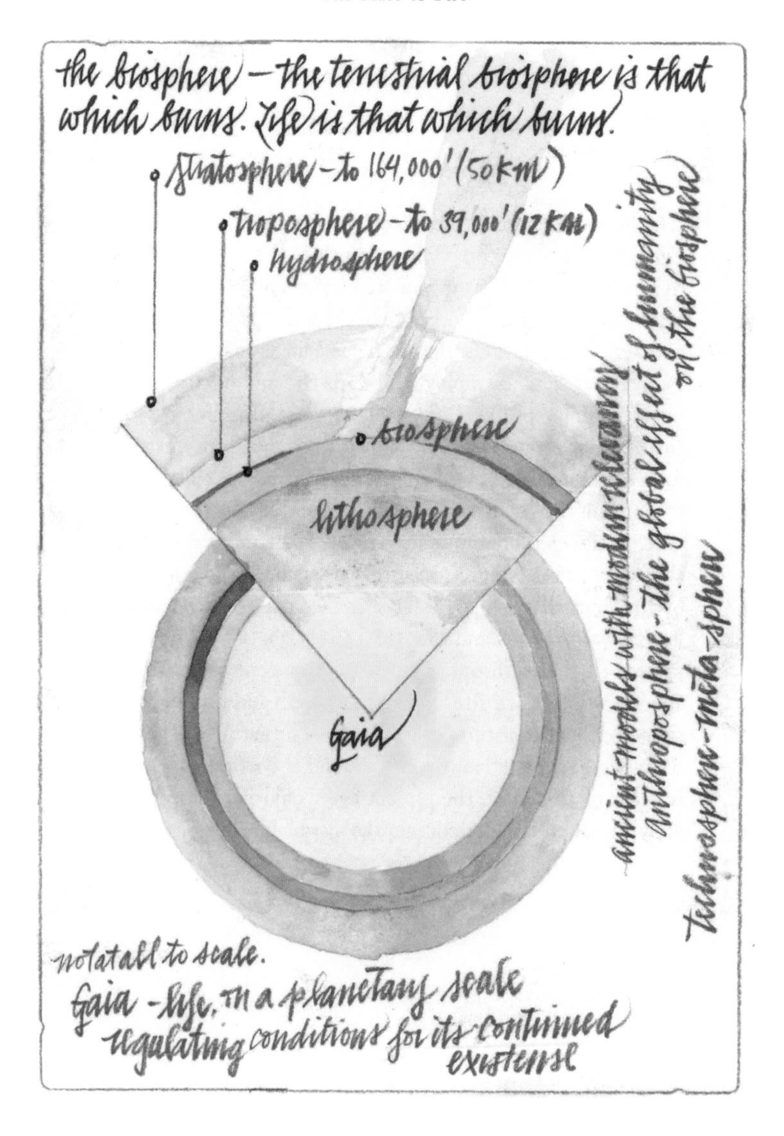
the biosphere – the terrestrial biosphere is that which burns. Life is that which burns.
stratosphere – to 164,000' (50 km)
troposphere – to 39,000' (12 km)
hydrosphere
biosphere
lithosphere
Gaia
ancient models with modern relevancy
anthroposphere – the global effect of humanity on the biosphere
technosphere – meta-sphere
not at all to scale.
Gaia – life on a planetary scale regulating conditions for its continued existence

Twenty-five hundred years before now, the pre-Socratic Greeks conceived of the four classic elements: air, earth, water, and fire. Applied to the globe, fire and life occupy the same slot. In 1979, the climate scientist James Lovelock proposed a theory of how the living world works, which he coined Gaia theory. Gaia is a holistic theory of the biosphere which proposes that life, as a global force, is a self-regulating and self-perpetuating system that creates planetary conditions convivial to its own continuance. The living system, Gaia, works through a process called cybernetics, the interplay of communication and control among subsystems in any living body.[12] Evidence of Gaia at work includes statistical analysis of microorganisms in the ocean, sequestering carbon and regulating atmospheric conditions; networked mycorrhiza throughout the world's topsoil, circulating nutrients and perpetuating fecundity; and land fire working as a restorative agent and as a process of disturbance that liberates otherwise inaccessible energy and resources into the ecosystem and across the biosphere.

*Fires rival volcanoes for atmospheric disturbance. The big ones can produce such quanities of smoke that for several months, the plume may be hundreds of miles wide, piercing the stratosphere to altitudes in excess of ten miles.*

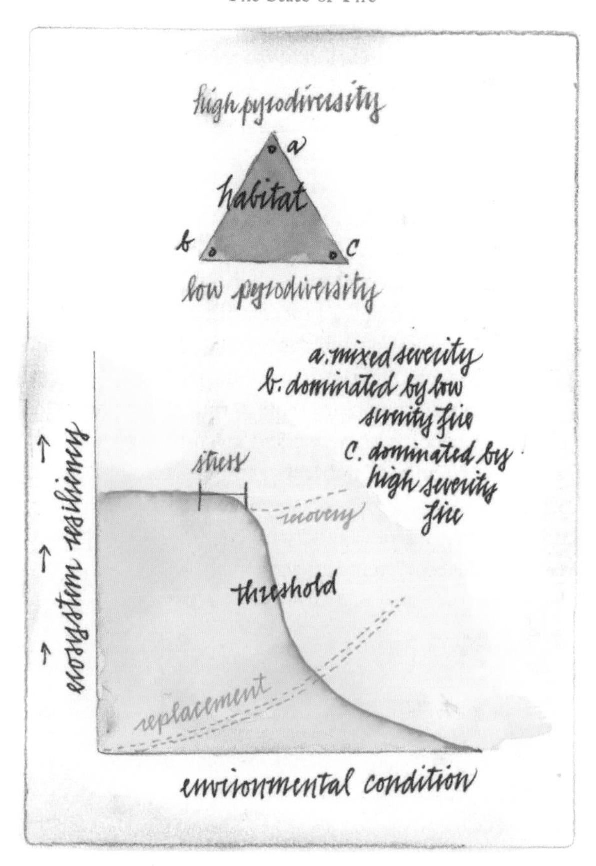


*Source (upper diagram)*: Adapted from G. Jones, J. Ayars, S. Parks, H. Chmura, S. Cushman, and J. Sanderlin, "Pyrodiversity in a Warming World: Research Challenges and Opportunities," *Current Landscape Ecology Reports* 7: 49–67, http://doi.org/10.1007/s40823-022-00075-6.

*Source (lower diagram)*: Adapted from S. M. Munson, S. C. Reed, J. Peñuelas, N. G. McDowell, and O. E. Sala, "Ecosystem Thresholds, Tipping Points, and Critical Transitions," *New Phytology* 218 (2018): 1315–17, https://doi.org/10.1111/nph.15145.

## 01.02 Pyrodiversity and Biodiversity

### *Resiliency and postfire ecological response*

Just as biodiversity is the measure of variability of species, ecosystem types, and genetic information, pyrodiversity is the measure of the variability of fire's return time, its patchiness on the land, and its burn severity. With greater pyrodiversity in any given landscape there is a corresponding increase in niche resources that become available, unlocked by fire. A niche resource is any exploitable opportunity, and with regard to what fire provides, a small list of these resources might include increased access to habitat, prey, sunlight, nutrient chemicals and fertilizers, and necessary temperature thresholds for particular fire-dependent processes to occur. Competition to utilize the novel resources that land fire provides is the basis for the empirical claim that pyrodiversity promotes biodiversity.[1] More types of fire mean more resource types that become available, which may mean that more forms of emergent adaptations present themselves as strategies for new species to specialize in making use of. Because of its scalable intensity and severity across the ecosystem, land fire may bestow a greater number of resource opportunities than any other large-scale disturbance event.[2] Greater pyrodiversity might not mean greater biodiversity everywhere, but in California, fire has been a major impetus for evolution. California is home to almost sixty-three hundred plant species and subspecies. That number is about one-quarter of the total number of plant taxa in all of North America, and California comprises only 4 percent of the continent's land mass. Of the total number of plant species in California, almost one-third (just over 2,150 species) are endemic to the state.[3]

## *Resiliency is the measure of intact recovery*

Resiliency is a measure of how an ecosystem and its populations of plants and animals recover after a disturbance. If a hilly terrain of oak woodland goes up in flames because of a moderately severe fire, the capacity for that environment to regenerate and welcome back the flora and fauna which call that habitat home is a measure of the ecosystem's resiliency. Resiliency can also be described as an ecosystem's response capacity, or its adaptive capacity to bounce back from damage. What bounces back is the identity of the ecosystem defined by its functionality, and that functionality is defined by the native biodiversity within the ecosystem.[4] Biodiversity defines ecosystem functions through relationships of life that govern energy production (what plants are present to form the basis of the food chain), how nutrients are cycled (what plant, animal, and fungal relationships are present across the food web), and how decomposition works (how organic matter moves through the environment).[5] If that oak woodland comes back quickly, it is highly resilient. Perhaps there is a good mix of old trees of many different hardwood species, and fire is known to regularly occur at the right frequency to keep fuel levels low. If the oak woodland doesn't come back, or comes back highly altered, its resiliency is low; perhaps invasive grasses are everywhere, and fire, either having come back too often or not at all, has left the trees vulnerable to insect and fungal blight. The capacity of the ecosystem's functionality to absorb a certain amount of stress from any given disturbance is referred to as its resiliency threshold.[6] The threshold represents a level of vulnerability beyond which the ecosystem's ability to maintain its own integrity is threatened. A weakened oak woodland that has seen too much fire because of human ignitions or has endured too much drought has a low resiliency threshold and a relatively higher chance of succumbing to ecotype change.

The key variable in any ecosystem that predicts its long-term resiliency and maximizes its adaptivity while minimizing its vulnerability is the relative quality and amount of its biodiversity before the disturbance event.[7] There are three types of biodiversity: (1) genetic biodiversity, or the variety of genetic information within individuals of a particular species; (2) species biodiversity, or the variety of types of species within a particular habitat or ecosystem; and (3) ecological biodiversity, or the variety of environments, habitat types, or ecosystems within a particular land area.[8] Transcending the threat of wildland fire, and far from being an academic abstraction, there is a direct through line from biodiversity to the well-being of human communities.[9] Biodiversity means healthy ecosystems, and healthy ecosystems mean the reliable delivery of ecosystem services, which range from pest control to clean water and normalized fire regimes. As has been the case with every human community, regardless of scale and since the beginning of time, the reliable delivery of ecosystem services means thriving human communities. In the era of climate change, defending biodiversity is among the highest priorities for shoring up the resiliency of human ecology across California and the biosphere.[10]

*One wildflower, staggering diversity*

The genus *Clarkia* contains forty-two species (among them farewell-to-spring), and thirty-six of those species are endemic to California. The diversity of this successful genus is a result of mutations, gene incompatibility (breeding system changes), drought and soil extremes (environmental stress), and pollinator diversity (reproductive strategy). Every endemic species of *Clarkia* was established in California millions of years before the establishment of California's regular, Mediterranean-type climate. These flowers are older than the shape of these mountains.

Elegant
Clarkia
Clarkia
unguiculata
Red
ribbons
Clarkia
concinna

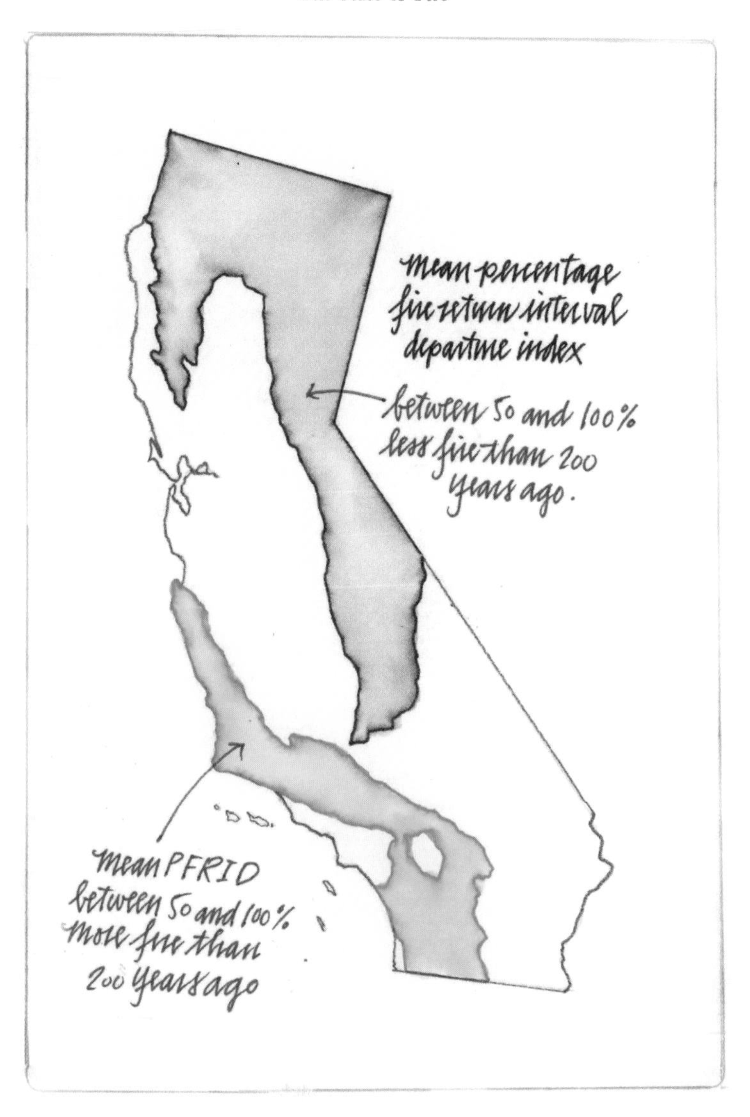
mean percentage
fire return interval
departure index
between 50 and 100%
less fire than 200
years ago.
mean PFRID
between 50 and 100%
more fire than
200 years ago

## *Fire in two Californias*

The fire return interval, or fire return index (FRI), is the length of time between fires in a normalized fire regime within a particular ecosystem. The percentage fire return interval departure (PFRID) is the difference, expressed as a percentage, between present FRIs and pre-Euro-American settlement estimations within the same ecosystem. Higher percentages mean there is too much fire as compared to what once was, and lower percentages mean there is less fire than there once was. Over the past century, because of fire exclusion policy and urban development, a significant trend of fire return deviation has developed in opposite directions between the north and the south of the state. Across Northern California, increases in the interval between low-intensity fires have led to a kind of fire debt, and across Southern California, decreases in the interval between high-intensity fires have led to a fire glut. Both scenarios threaten human security and make landscapes less resilient. These substantial departures from past timing for fire have resulted in deformities in ecological composition, structure, and function.[11] Midmontane forests and woodlands expect to see low-severity, frequent fires, but for a century they have largely only experienced high-severity, infrequent fires. Meanwhile, chaparral always burns as a crown fire, and it only wants to do so approximately every hundred years.[12] Chaparral covers more than seven million acres of land area in California, or 7 percent of the state, and supports almost one-quarter of the state's biodiversity, half of which is endemic to chaparral ecosystems.[13]

## 01.03 A Tragedy of Misconceptions

### *When the wilderness is not wild*

When the Americans settled California in droves during the latter half of the nineteenth century, many brought with them a fundamental misinterpretation of the core character of the land. California's "wilderness frontier" was not wilderness, and it was not a frontier. Most American settlers had little idea of what they were looking at and carried only a vision of what they thought they could turn it into. Worse, they often invented what they wanted to see in California's landscape and conceived of it as primeval and ripe for wholesale transformation.[1] This myopic view of California as a wild Eden led to an extractive land ethic that ignored the endemic adaptive cycles supporting the land's ecosystems since the earliest days of the Holocene. Worse, it ignored the ancient effect of people on the land, who worked with a sophisticated knowledge of best practices for stewarding ecological health. One of the primary tools used by so many tribal sovereignties across California was the regular application of fire, often across large swaths of land. To the settler, Indigenous fire in California was—and often continues to be—regarded as deleterious to the landscape. The tragic miscalculation was thinking that the land needed to be saved from fire when in fact the land thirsted for it.

To understand modern fire policy in California requires an analysis of California's settler attitude toward the more-than-human world. This attitude relies heavily on the invented concepts embedded in the words *nature* and *wilderness*. To confront the many injustices that are still being perpetuated to theoretically protect what these words represent from what is conceived of in the settler mind as the deforming hand of culture, it is important to imagine that these words don't exist as objective states but rather as political designations. Nearly all habitat space of precontact California was stewarded for some kind of anthropogenic purpose, through disturbance regimes that were complex and massive in scope.[2] If California has ever represented nature in balance—a natural wilderness in a steady state, as many nineteenth-century colonizers imagined it—that balance existed not despite but because of regular perturbations within the state's ecosystems.[3] The relationship between fire, biodiversity, and human stewardship in precontact Holocene California that has existed for tens of thousands of years may be so complete that it forms a conceptual tripod that cannot stand without all three legs.[4]

California's dynamic ecology led to hundreds of pyrodiverse human economies that persisted for thousands of years.[5] When tribal groups decided (decide, and will decide) when and how

often to set fires, their decisions were (and are) engineered to manage a particular mix of economic resources based on fuel sources and ignitions, climatic and weather conditions, and the landscape patterns of previous fires.[6] Enduring and thriving through millennia of dynamic climatic fluctuations, from centuries of drought through decades of deluge, the Indigenous people of California used fire to construct landscape-wide regimes of resilient pyrocultural food production. European history has often presumed the use of a single methodology of agriculture whose origins stretch back to Babylon. California pyrocultural technology wasn't a precursor to European-style agriculture but more likely a conscious eschewing of irrigation and till-based farming.[7] Pyrocultural food production reflected a variety of cultural traditions in what can be called commensal reciprocities, or kinship economies, that acted in accordance with the living rhythms of the local ecology.[8]

If it is true that what was untouched, unworked, and untended in California prior to American settlement fades to a diminishingly small bit of unknown geography, what then is natural? What is wild? Might words like *natural* and *wild*, used from a settler colonial perspective and applied to thoroughly stewarded land, be something like a perpetuation of injustice? Associated words, such as *pristine* land or, even stranger, *virgin* land, only make sense in terms of purity, sexualization, or domination. The concept of wilderness may imply degradation and neglect in the context of what can be thought of as an Indigenous critique of modern land policy. Wilderness represents a fallen state into which the land descends in the absence of work not done and relationships not tended to, as expressed by Sierra Miwok elder James Rust, quoted by M. Kat Anderson: "The white man sure ruined this country, it's turned back to wilderness."[9] Wilderness holds a kind of romance for the colonizing mind of the twentieth century,

In the old stories of Coyote, when he was the chief of the people, when animals were people, before the land and the people and the animals were fixed, sacred time flowed in a circle. Sacred time exists within circles stacked inside of larger circles, and it does not make sense to conceive of history as events that happened, but rather as something that is unfolding now, in the living stories of consequence between beings. There is nothing in the world but the quality of relationships between beings.

explicitly and eloquently laid out by Wallace Stegner in his 1960 letter to the US Congress in defense of what would pass as the Wilderness Act four years later. In that letter, Stegner describes a "geography of hope" that wilderness bestows, a kind of sanity that modern development on that landscape would threaten.[10] The Indigenous critique might counter that this policy is an invention that denies ecological reality. Preserving that thing called nature means imposing a stasis on the systems that deliver any given ecosystem's vitality, effectively condemning so many such systems to overgrowth, stress, infirmity, invasivity, and disease. On the state and federal levels, the future of colonial policies in California that are predicated on the concepts of nature and wilderness will hinge on the ability to articulate the intersection between justice and ecology in terms of land management practice, tradition ecological knowledge, and fire science.

"Am I ever going to learn?" Coyote repeated again and again to himself. "I just keep making mistakes."[11]

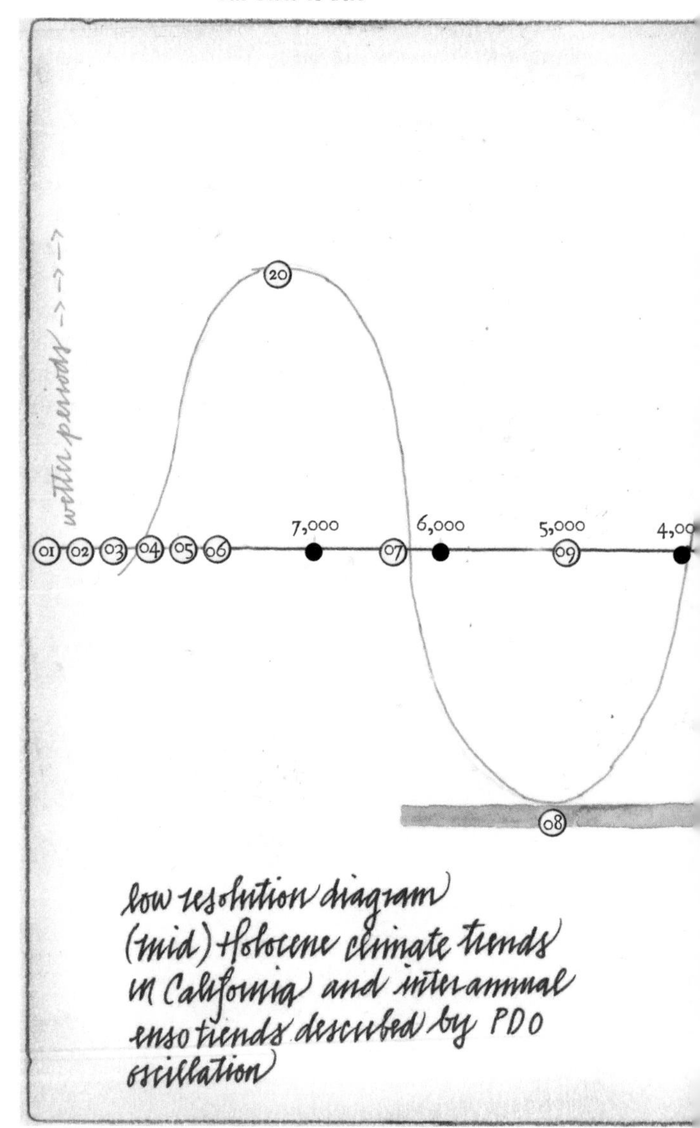
wetter periods -> -> ->
20
7,000
6,000
5,000
4,00
01
02
03
04
05
06
07
09
08
low resolution diagram
(mid) Holocene climate trends
in California and interannual
enso trends described by PDO
oscillation

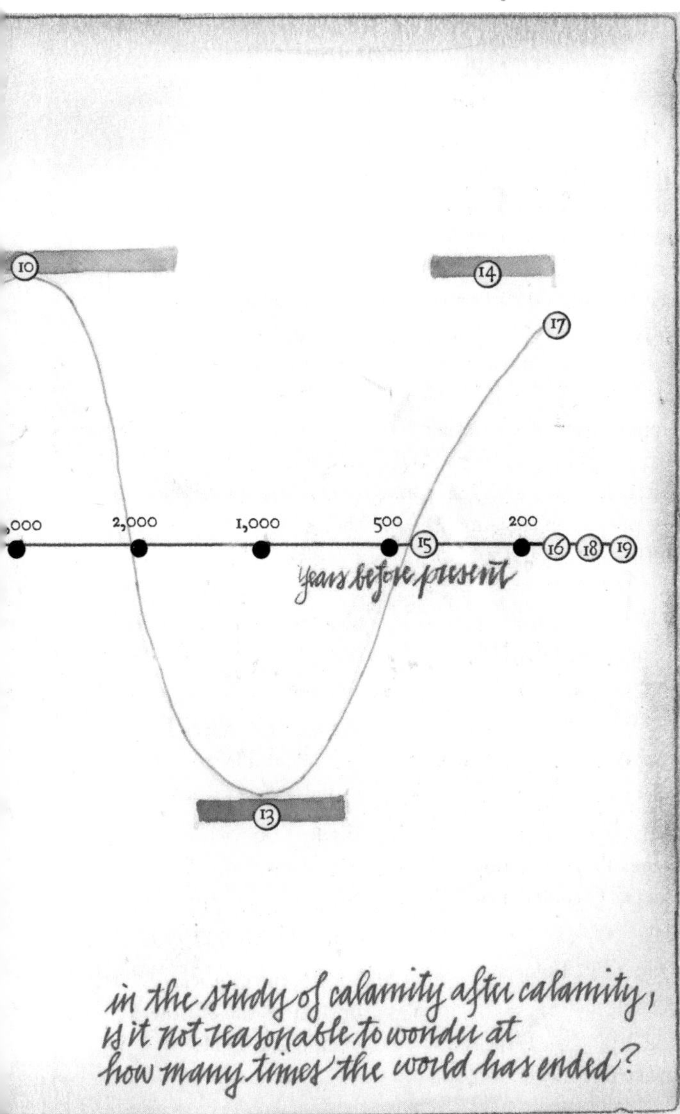
10
14
17
,000
2,000
1,000
500
200
15
16
18
19
Years before present
13
in the study of calamity after calamity,
is it not reasonable to wonder at
how many times the world has ended?

## *California climate trends and archaeology through the Holocene*

Earth's climate is breaking down at an unprecedented rate due to anthropogenic global warming. But we can learn from California's climate history to understand the changes that are unfolding now. Notable events over the past ten thousand years include the Altitherm, or the Mid-Holocene climatic optimum, when the San Francisco Bay formed and when many modern tribes, as we know them today, may have established their current distributions.[12] During this time, which included a major, centuries-long drought and must have been a time of huge fires, California retained its entire complement of species biodiversity, meaning that no plant or animal seems to have succumbed to extinction.[13] Also notable is the Little Ice Age, which may have resulted from the genocide of native peoples in the Americas and the centuries-long pause in cultural fire on the land.[14]

01. 20,000 YBP (years before present), Last Glacial Maximum; carbon 185 PPM; 6° C cooler than today.
02. 15,000 YBP, early archaeological evidence.
03. 12,900–11,700 YBP, Younger Dryas Cooling Event; 5° C cooler than today.
04. 11,700 YBP, Holocene begins.
05. 8,300 YBP, San Francisco Estuary forms over 700,000 acres (1,000 square miles with 48 square miles of wetland).
06. 8,000 YBP, Silver Lake, last of the Pleistocene lakes in the Mojave, dries up.
07. 6,500 YBP, San Francisco Bay forms.
08. 7,500–4,000 YBP; 3°C warmer than today.

Mid-Holocene climatic optimum—the Altitherm; carbon 260 PPM.[15]

6,500–4,500 YBP mid-Holocene drought.[16]

09. 5,200 YBP, West Berkeley shell mound.

5,000 YBP, coastal upwelling and development of summer fog begins.[17]

5,000 YBP, modern distribution of California's forests.

10. 4,000–2,000 YBP, neoglaciation.

11. 3,730 YBP, Mono Lake high point (6,500' elevation).

12. 3,260 YBP, Silver Lake in the Mojave re-forms.

13. 1,500–500 YBP, Medieval Drought/Medieval Warm Period.

Two megadroughts separated by a wet period 1100–1200 YBP.[18]

3° C warmer than today.

Water temperatures in the eastern Pacific 2° C cooler than today.

14. 400 YBP Little Ice Age, 16th–19th centuries.

Cyclical lows in solar radiation.

Native American genocide.

15. 1605, megaflood in the Central Valley.

16. 1861, megaflood in the Central Valley.

17. Climate trends toward increasing instances of dry–wet climate knockout (also called weather whiplash, where one year may be abnormally dry and the following year or series of years may be equally abnormally wet.)[19]

18. Over the past 560 years, most megafire years have followed unusually wet periods.[20]

19. The time since the gold rush has been an aberrantly wet period, with the longest drought lasting only six years. In the past two millennia, droughts lasting for decades occurred at intervals of every 50–90 years.[21]

20. The sine wave depicts a generalized representation of the 1,500-year oscillation between dry and wet millennia in California during the Holocene.[22]

# 01.04 Cultural Fire on the Land

## *Traditional ecological knowledge*

For at least ten thousand years, humanity became largely naturalized in the California landscape. Indigenous science and successful land management practices, including strategies for using prescriptive fire, were characteristic of a process that has been called ecofunctionalism.[1] This way of tending nature sustained hundreds of thousands of people. Its underlying land ethic was one of reciprocity, where processes (such as cultural fire) that propagate biodiversity and soil fecundity were managed to the mutual benefit of both the human and the more-than-human world. A land ethic of reciprocity implies a coevolved relationship in which the land and the people need each other. Whether Paleo-Indians began with this land ethic or it was an emergent philosophy remains contested. There is a common theory that in the early Holocene, humans participated in exploitative practices that do not seem to have aligned with the theory of reciprocity. Most notably, the first peoples of California participated in an overkill of Pleistocene megafauna that resulted in the extinction of scores of species.[2] Whatever the case, humans became stewards of the land (and surely then were stewarded themselves), accounting and responsible for what may have been most of California's ecological productivity for possibly tens of thousands of years.

From the application of regular, low-severity fire across millions of acres every year, to the seasonal harvesting of hundreds of different plant species for ethnobotanical reasons—whether dietary or medicinal—these cultural technologies are still evident in the structure, configuration, and distribution of California's ecosystems today. It is estimated that before 1800, between 2 and 12 percent (approximately 12 million acres) of the entire land area of the California Floristic Province was intentionally burned on an annual basis by cultural fire. (For comparison, recent averages include an estimated annual burning of about 2 million acres.[3]) Whether these fires were large enough and widespread enough to prevent megafires is still debated.[4] This land management resulted in pyrodiverse foods: nuts, seeds, grains, greens, fruits, bulbs, corms, rhizomes, taproots, and tubers. And the methods of producing these foods was just as diverse: burning and fire management, seasonal pruning, digging and in some cases irrigation, weeding and clearing debris, transplanting, coppicing, and harrowing.[5] Indigenous economies and ecosystems benefited from fire-cleared landscape in other ways too: easier foot travel and gathering, which resulted in a wider distribution of seed varietals; the destruction of diseased plants and pathogenic pests; and the enhancement of nutrients in the soils.

Over thousands of years, land management of this kind produces three kinds of ecological effects: genetic modification, wider dispersal patterns of plants and animals, and habitat modification.[6] Through human selection, the genotypes of individual plant species can be altered in a relatively short period of time to respond to cultivation. For example, devil's claw (*Proboscidea parviflora*) is cultivated by the Paiute; there is a domesticated and wild genome of the same species.[7] The altering of the historical location and range of plant types seems to have been a regular occurrence, and examples might include common but anomalous groves of Northern California black walnut (*Juglans hindsii*) in the East Bay and the Sacramento River Valley.[8] Habitat alteration also seems to have been a common and expected result of the use of fire in land tending, including the maintenance and expansion of coastal prairies, the preservation of montane meadows, the shaping of oak savannas, and the promotion of sugar pine (*Pinus lambertiana*) in the conifer forests of the Sierra Nevada.[9]

A principle that is generally anathema to the standard model of tilling and agricultural practice but core to traditional ecological knowledge is scientifically named the *intermediate disturbance hypothesis*. Fire disrupts the ecosystem within parameters of resiliency (adaptivity of the species present, moisture, fuel load, etc.) that will determine the quality of the ecosystem's succession, or its structure upon regrowth.[10] For most terrestrial habitats in California, intermediate disturbances brought by cultural burning result in higher species richness.[11] For example, low-intensity, regularly applied fire influences the black oak/ponderosa pine forest—both the individual plants and the plant communities—of the midmontane Sierra Nevada in several beneficial ways. It increases mycelial activity in the soil, promotes epicormic growth in the oaks to produce more acorns, inhibits the spread of pests and pathogens, provides fertilizing nutrients for understory grasses and forbs, and eliminates brush to prevent future crown fire and prevent shade-loving trees, such as fir, from colonizing and overcrowding groves.[12]

Two keystone species—the western scrub jay and the acorn woodpecker, both intelligent and sociable birds—are largely responsible for determining how the oak woodland looks and functions. Jays and woodpeckers rely on acorns, just as the oak trees rely on the birds for distribution. Scrub jays have bills that are uniquely adapted to pull, transport, and open acorns, and although many species cache nuts, acorn woodpeckers are the only species to drill an individual hole for each nut they harvest. In the fall, a flock of scrub jays might collectively remove four hundred acorns an hour from a single tree, and some acorn granaries (trees riddled with woodpecker holes) can contain fifty thousand acorns. Because jays prefer bigger acorns than woodpeckers, the two species don't directly compete, and the oak tree maximizes the spread of its acorns through the environment.

ohlone food – fire food of the Bay Area
(incomplete)

tansy-mustard seeds
sage (chia) seeds
evening primrose seeds
clarkia seeds
red maid seeds
brome seeds

acorns
seeds
roots
nuts
berries

autumn
blackoak
tanoak
live oak
Valleyoak
holly leaf
cherrypits

hazelnuts
fannel nuts
blackwalnuts

Buckeye seed
must be:
1. roasted
2. peeled
3. mashed
4. leached (for 18 hours)
poison: prussic acid

cattail roots
brodiaea bulbs
mariposa lily bulbs
soaproot bulbs
soaproot

grapes
currants
goose berries
salal berries
elder berries
thimble berries
toyon berries
madrone berries.
huckle berries
manzanita berries

Greypine seeds
winter mushrooms

elk, deer, antelope
grass hoppers
freshwater fish

greens
clover
poppy
tansy-mustard
melic grass
rooteh
mule ear shoots
cow parsnip shoots

alum roots (young leaves)
columbine
milkweed
larkspur

fire-stewarded foods of the Bay Area

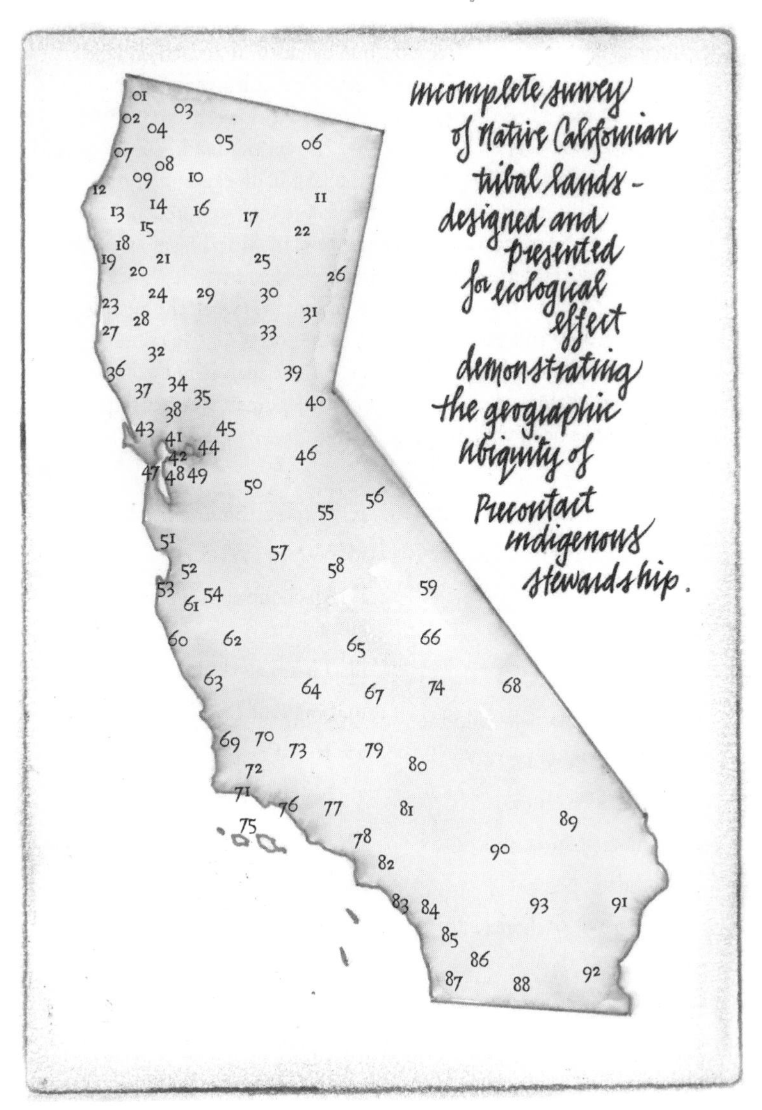
incomplete survey
of Native Californian
tribal lands -
designed and
presented
for ecological
effect
demonstrating
the geographic
ubiquity of
Precontact
indigenous
stewardship.
01 02 03 04 05 06 07 08 09 10 11 12 13 14 15 16 17 18 19 20 21 22 23 24 25 26 27 28 29 30 31 32 33 34 35 36 37 38 39 40 41 42 43 44 45 46 47 48 49 50 51 52 53 54 55 56 57 58 59 60 61 62 63 64 65 66 67 68 69 70 71 72 73 74 75 76 77 78 79 80 81 82 83 84 85 86 87 88 89 90 91 92 93

This map shows a generalized outline of cultural and tribal lands. This book often refers to Native Californian cultures collectively, in order to make broad arguments about traditional ecological knowledge, landscape alterations, and fire on the land. But it is important to note that there are thousands of existing and nuanced traditions in California, and just as many approaches to the practice of Indigenous pyrogenic stewardship. This map is therefore low resolution and incomplete by necessity. It shows approximate tribal regions alongside endonyms (what the people call themselves) and ethnonyms (names imposed on the tribes—listed here in parentheses). The intent of the map is not ethnographic but ecological, meant to show that human stewardship in California has been ubiquitous for a long time.

01. Tolowa Dee-ni (Tolowa)
02. Oohl (Yurok)
03. Araar (Karuk)
04. Huil'kut (Chilula)
05. Kahosadi (Shasta)
06. Moatkni Maklaks (Modoc)
07. Ku'wil (Wiyot)
08. Natinixwe (Hupa)
09. C'imar (Chimariko)
10. Nongatl (Nongatl)
11. Achomawi (Achomawi)
12. Ni'ekeni (Mattole)
13. Hoilkut (Whilkut)
14. Lassik (Lassik)
15. Ukomno'om (Valley Yuki)
16. Wintu (Wintu)
17. Mechoopda (Mechoopda Maidu)
18. Tlokyhan (Cahto)
19. Sinkyone (Sinkyone)
20. Kinist'ee (Wailaki)
21. Nomlaki (Nomlaki)
22. Atsugewi (Atsugewi)
23 Kahp-Bahl-Xee-Kaw-nuu (Northern Pomo)
24. Tceefoka (Northeast Pomo)
25. Yana (Yana)
26. Cui Ui Ticutta (Northern Paiute)

27. (Central Pomo)*
28. (Eastern Pomo)*
29. Cachil Dene (Patwin)
30. Garii (Yahi)
31. Maidu (Maidu)
32. Elem (Southeastern Pomo)
33. KonKow (Concow)
34. Hotsa-ho (Lake Miwok)
35. Klestal Dehe (Patwin)
36. Kashia (Kashaya Pomo)
37. (Southern Pomo)*
38. Micewal (Wappo)
39. Nisenan (Nisenan)
40. Waashiw (Washo)
41. Suisunes (Suisun Patwin)
42. Karkin (Karkin)
43. Micha-ko (Coast Miwok)
44. (Bay Miwok)*
45. Tumtum (Plains Miwok)
46. (Sierra Miwok)*
47. Ramaytush (Ramaytush)
48. Chochenyo (Chochenyo)
49. Tamyen (Tamyen)
50. Chulamni (Northern Yokut)
51. Awaswas (Awaswas)
52. Mutsun (Mutsun)
53. Rumsen (Rumsen)
54. Chalon (Chalon)
55. Chukchansi (Foothill Yokut)
56. Kucadikadi (Mono)
57. Tachi (Central Yokut)
58. Wukchumni (Kaweah River Yokut)
59. Nuumu (Owens Valley Paiute)
60. Esselen (Esselen)
61. Quinau (Salinan)
62. Cholam (Salinan)
63. (Obispeno Chumash)*
64. Tuilmni (Buena Vista Yokut)
65. Yaudanchi (Tule River Yokut)
66. Newe (Owens Valley Shoshone)
67. Paleayani (Poso Creek Yokut)
68. Numu Tumpisattsi (Timbisha)
69. (Purisemo Chumash)*
70. (Interior Chumash)*
71. (Barnareno Chumash)*
72. (Yneseno Chumash)*

73. (Emidiegeno Chumash)*
74. Tubatulabay (Tubatulabay)
75. (Island Chumash)*
76. (Ventureno Chumash)*
77. Tatavium (Tatavium)
78. Kitanemuk (Kitanemuk)
79. Nuwu Kwatsaan (Nuwu Kwatsaan)
80. Vanyume (Vanyume)
81. Kitanemuk (Kitanemuk)
82. Tongva (Tongva)
87. Kamia (Kumeyaay)
88. Titpay (Kumeyaay)
89. Inuwuwu (Chemehuevi)
90. Taagtam (Serrano)
91. Xalychidom (Haichidoma)
92. Kwatsaan (Kwatsaan)
93. Ivilyugaletem (Cahuilla)

*Precontact endonym unknown

# 01.05 Colonial Conflagration

## *Genocide and ecocide*

Although Spain established the mission system of the Californias in 1760, it arguably wasn't until Junipero Serra arrived in San Diego on the first day of July 1769 that a conquering European power, in the form of the Catholic church, arrived with devastating effect in California.[1] Before his death in 1784, as the president of the missions of the Californias, Serra would oversee the conscription and death of many tens of thousands of Indigenous Californians.[2] In 1773, with the arrival of the first several hundred cows to supply the mission system with food—along with invasive pasture grass, Serra's wine-grape, and other European agricultural staples—California's land use and basis for food production would be forever transformed.[3] With the first prohibitions against cultural fires by Spanish colonial administrators in 1793—a move that presumably was about saving grassland for pasture, but in effect was to further enslave the native population—California's landscape would also be forever transformed.[4] Serra's arrival prefigured the coming of both genocide and megafire to California. But present-day, out-of-control fires are also symbolic of colonialism itself, which is transformative in the most calamitous way.

The symbolic megafire of colonialism still rages today. It rages in state policies that are only now catching up with science, democratic will, and a resurgent and mobilized Indigenous community. It rages today in a California citizenry that remains largely ignorant of the historical injustices perpetuated by its government and is bewildered by the existential threats of the climate and biodiversity crises.

California and national fire policy timeline and significant western fires, 1849–1999, with commentary:

1846—California genocide: between 1846 and 1873, more than sixteen thousand native peoples are brutally killed in a state- and federally sponsored campaign of violence extending across 370 individual massacre events in California. The population of Indigenous people drops due to disease and genocide from approximately 150,000 to 16,000.[5]

1849—The city of San Francisco burns seven times in sixteen months

*What a clumsy creature this city seems to be—sparking into existence, constructed from the bones of old-growth redwood, California's first city explodes in population and burns, again and again*

1860—Creation of the California Geological Survey, which will eventually direct continuing programs including the Burned Watershed Geohazards Program, the Forest and Watershed Geology Program, and the Regional Geologic and Landslide Mapping Program.

1864—The US government gifts the Yosemite Grant and Mariposa Grove, the ancestral home of the Ahwahneechee people, to the State of California. Pre-contact, Indigenous fire application maintained ecosystems in Yosemite Valley for food production with only half as many trees as there are today; trees averaged 20 percent larger when compared to current populations.[6]

1868—Founding of the University of California, Berkeley, in Strawberry Canyon, the ancestral lands of the Southern Huchiun Chochenyo people, known now as the Muwekma Ohlone tribe.

Today Cal is home to UC Berkeley's Center for Fire Research and Outreach, and the Stephens Fire Science Laboratory at the university's Rausser College of Natural Resources. The first class of forestry students from UC Berkeley graduates in 1914.

1885—Establishment of the California Board of Forestry, which will become CAL FIRE, with the dual purpose of protecting "merchantable" timber and protecting landscapes from wildland fire.[7]

1905—The Forest Protection Act establishes the US Forest Service (USFS), which assumes control of forest reserves. The first director of USFS is Gifford Pinchot, who serves until 1909, and despite being trained as a forester in Europe, has a relatively nuanced vision of the regenerative power of land fire.[8]

1910—The Big Burn, the largest wildfire in US history, reaches three million acres in forty-eight hours on the Idaho-Montana border. This event galvanizes the USFS as an organization with a priority mission to prevent and battle every wildfire.[9]

1911—*Systematic Fire Protection in the California Forests* by Coert DuBois is published. This influential publication institutionalizes the scientific case, albeit a dated one, for the policy paradigm that fire exclusion is forest protection.

1930—The founding of the Civilian Conservation Corps and the construction of the Ponderosa Way, an eight-hundred-mile-long firebreak that stretches from Mount Shasta to Bakersfield.

*The Ponderosa Way—the lost Depression-era firewall of California, probably could not have saved Paradise from the Camp Fire—the Ponderosa Way could do very little to stop twenty-first-century flashover wrought by climate-fueled megafire.*

1932—The Matilija Canyon Fire rips through 220,000 acres of Ventura County, a prelude to the massive blaze that would visit the same region as the Thomas megafire almost a hundred years later.

1942—The Wartime Ad Council creates the first national program on fire prevention. Two years later, the Smokey Bear campaign initiates arguably the most successful ad campaign of all time. Smokey cements the pervasive myths that (1) wildland fire is bad, and is irredeemably so, and (2) individuals, not corporate interests, are the only preventive force to stop wildfires from beginning.

1945—The California Disaster Act passes, enabling the governor to declare a state of emergency. (Intended to fund and coordinate statewide agencies against fire and war. Its successor, the 1970 California Emergency Services Act, is invoked in Governor Newsom's statewide 2020 COVID response.[10]) The Forest Practice Act, also passed in 1945, mandates that all private foresters must inform the state before harvest.

1949—Publication of Aldo Leopold's *Sand County Almanac*, a classic of nature conservation that popularizes two core ideas: "thinking like a mountain"—a deep-time concept that heavily influences the environmental movement of the 1960s and '70s—and the "land ethic," which lucidly provides a moral basis for conserving the natural world in modern times. Leopold's book influences the Tall Timbers organization, an annual national symposium on fire ecology that launches in 1962.[11]

> *The land ethic simply enlarges the boundaries of the community to include soils, waters, plants, and animals, or collectively: the land.*
>
> *—Aldo Leopold*

1956—Militarizing the firefighting effort in California, Operation Firestop sees the USFS and California Department of Forestry use war vehicles and technology for the first time.[12] Aerial tankers drop retardant for the first time on the Inaja Fire (San Diego County), which burns forty thousand acres and kills eleven people.

1961—Elite fire crews called "hotshots" are deployed for the first time to fight the Bel Air Fire, which destroys nearly five hundred homes in the wealthy Los Angeles neighborhood.[13]

1964—Passage of the Wilderness Act, which establishes federal wilderness areas—the US government's most stringent land protection designation—perpetuating assumptions that leaving land parcels without regular stewardship or development is land protection. The act is supported by and perhaps passed because of California author Wallace Stegner, who in a letter to Congress described the wilderness as "the geography of hope."[14]

*We simply need that wild country available to us, even if we never do more than drive to its edge and look in. For it can be a means of reassuring ourselves of our sanity as creatures, a part of the geography of hope.*

*—Wallace Stegner*

1970—Modern California's largest fire to date—the Laguna Fire complex, which blackens 577,000 acres of Los Angeles County—restructures the state's response to fire emergencies with an interagency program called Firescope, the template for today's National Incident Management System.[15] The next ten years see other massive fires in Southern California, including the 1977 Southern California fire siege (over 500,000 acres) and the destructive 1980 Panorama Fire (29,000 acres and more than three hundred buildings destroyed, four dead).

1980—The Vegetation Management Program is passed by the state legislature, instituting a prescribed burn of twenty-five thousand acres of Southern California shrubland for fire prevention. The 1980s see many fires, including the Indian Fire (1980; 28,000 acres, Orange County), the Gypsum Canyon Fire (1982; 17,000 acres, Orange County), and the Mount Baldy Fire (1984; 15,000 acres, San Bernardino County). Statewide firefighting agencies continue to consolidate under the aegis of the California Emergency Management Agency (1986).

1990—The '90s kick off with the Painted Cave Fire (5,000 acres, Santa Barbara County) and then the 1991 Tunnel Fire (1,500 acres, Alameda County), called the Oakland Hills Firestorm, killing twenty-five people. Several other fires scorch California's backcountry and wildlife-urban interface, but the decadal average remains under 350,000 acres—an average that would be destroyed in the following decade.[16] In 1999, the California Department of Forestry becomes the California Department of Forestry and Fire Protection (CAL FIRE).

# Part One: Fire History

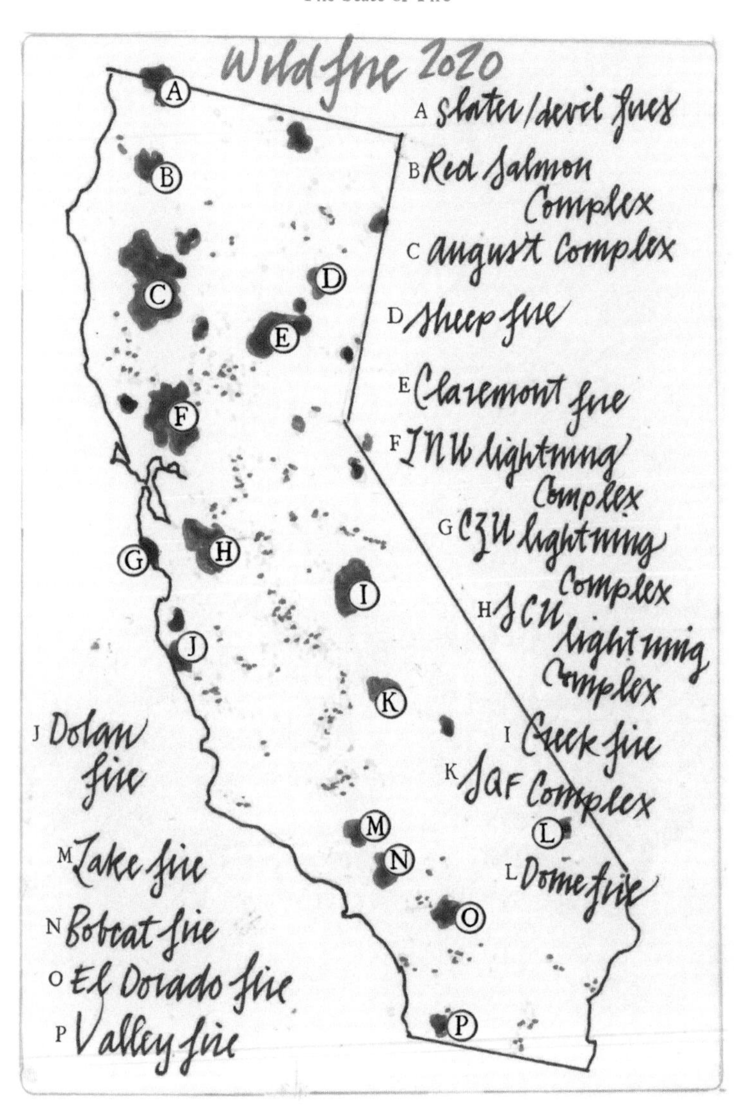
Wild fire 2020
A Slater/devil fires
B Red Salmon Complex
C August Complex
D Sheep fire
E Claremont fire
F LNU lightning Complex
G CZU lightning Complex
H SCU lightning Complex
I Creek fire
K SQF Complex
L Dome fire
J Dolan fire
M Lake fire
N Bobcat fire
O El Dorado fire
P Valley fire

# 01.06 The Coming of Modern Megafire

## *Why and how now is different*

There was always going to be a period of reckoning—with California's colonial legacy, with the state's history of fire management, with the practices of extractive industries, with our patterns of land development—and in the past twenty years it has arrived. California has entered an era of megafire. In accordance with the National Interagency Fire Center, the word *megafire* refers to any fire that is larger than one hundred thousand acres (156 square miles).[1] Eighteen of the twenty largest wildfires in the past two hundred years have occurred since the year 2003.[2]

The largest wildfire event in the past two hundred years was made up of thirty-eight smaller, lightning-sparked fires that converged to rage through and spill out of the Mendocino National Forest in the summer of 2020. Called the August Complex Fire, or the August 2020 California Lightning Siege, the combined fire complex grew to scorch more than one million acres (1,563 square miles), making it the first so-called gigafire in California history. The year 2020 turned out to become the biggest wildfire year ever for the state, with a total of almost ten thousand fires having blackened over 4.4 million acres, nearly 4 percent of the state. Although the sheer number of acres burned indicates a possible return to the quantity of acres burned under Indigenous fire regimes before Spanish and American colonization, the fire on the land may be too dramatic and severe.

one thousand reasons why the fire had to come
one thousand notes sung in ash,
drawn in tears
and burning mud
one thousand teachings
that feel more bitter
than time
One thousand roads
that were always going
to lead to this.
conclusion

To make matters even more concerning, these fires aren't influenced by local forces alone. Climate breakdown by way of anthropogenic global warming, and the ways that global forces intensify local problems, are core to emerging fire patterns in California. Ecological science indicates that as temperatures rise, precipitation patterns change, including decreased snowpack in California's mountains and a higher probability of statewide drought.[3] Combined with fire exclusion policy and the advent of the timber industry, more rigorous growth within forests due to a warmer climate has increased the statewide fuel load in conifer forests, as well as the connectedness of that fuel inside new fire corridors.[4] Drought-stressed arboreal habitats are increasingly vulnerable to fungal infection and beetle infestation, causing massive tree mortality events that raise the probability of fire.[5] The ravages of invasive plant species, habitat fragmentation and loss, biological pollution and dysfunction, and the depletion of resource stores have radically altered burn patterns across the California Floristic Province and the deserts of California.[6] The expansion of urban development, residential neighborhoods, and electrical infrastructure in the fire-prone landscapes of the wildland–urban interface (WUI, pronounced *woo-ee* by fire professionals)—not to mention the rise in arson events—has led to a dramatic rise in human ignitions and the spread of wildfire through human communities.[7]

Three primary factors that describe societal and climatic conditions in the early twentieth century set the stage for the era of megafire in the first two decades of the twenty-first.[8] The first factor: thanks to the invention of the internal combustion engine, the advent of the automobile, the expansion of the rail system through California's woodlands, and the increased opportunities for wildfire ignition, California experienced more fires in the 1920s than in any other decade in the twentieth century.[9] The second: to protect timber commodities, foresting companies of the early

1900s led the charge to keep fire out of California forests, and the USFS codified the request with the 1935 "10 a.m. policy," which mandated that all fires be extinguished by 10 a.m. the day after their discovery.[10] And the third: the century between 1850 and 1950 saw the fewest drought events in California in over two thousand years.[11] The result of a century of fire exclusion, coupled with the urban forests of human communities, led Governor Brown to say in 1962 that there were more trees, more tree cover, and more forested land in California than ever before.[12] California was covered in smaller, younger, overcrowded, vulnerable, and stressed trees—not in healthy ancient forests. With the decline of the timber industry in California in the 1970s, fire suppression strategies began to evolve into fire management strategies, as the 10 a.m. policy was slowly replaced with letting big fires rage and managing their effects by modeling and thus anticipating and potentially mitigating their behavior.[13]

Year after year of endless megafire is not inevitable or an eternal state of affairs. The causes of this decades-long march of megafire incidents are recent. Depending on how the current ecological bottleneck is resolved, the increasing emergence of megafires as unceasing, yearly events may be temporary. This chapter does not describe a situation without hope of relief from endless, repetitive fire. Developing and popularizing a nuanced understanding of wildfire's perennial and dynamic character will lead to the development of a road map through many perilous years to come. California is at a crossroads, and what choices are made in the near future regarding the issue of fire will be remembered by future generations as perhaps the most important that the people of the state ever had to collectively make.

fireweed, Chamerion angustifolium
a fire-following flower

Inside the stories told of the world,
there is stress and there is resiliency
inside the narrative systems that
must assemble themselves for the
Vision to cohere. There is always a
wounding and a healing,
often at the same
moment.

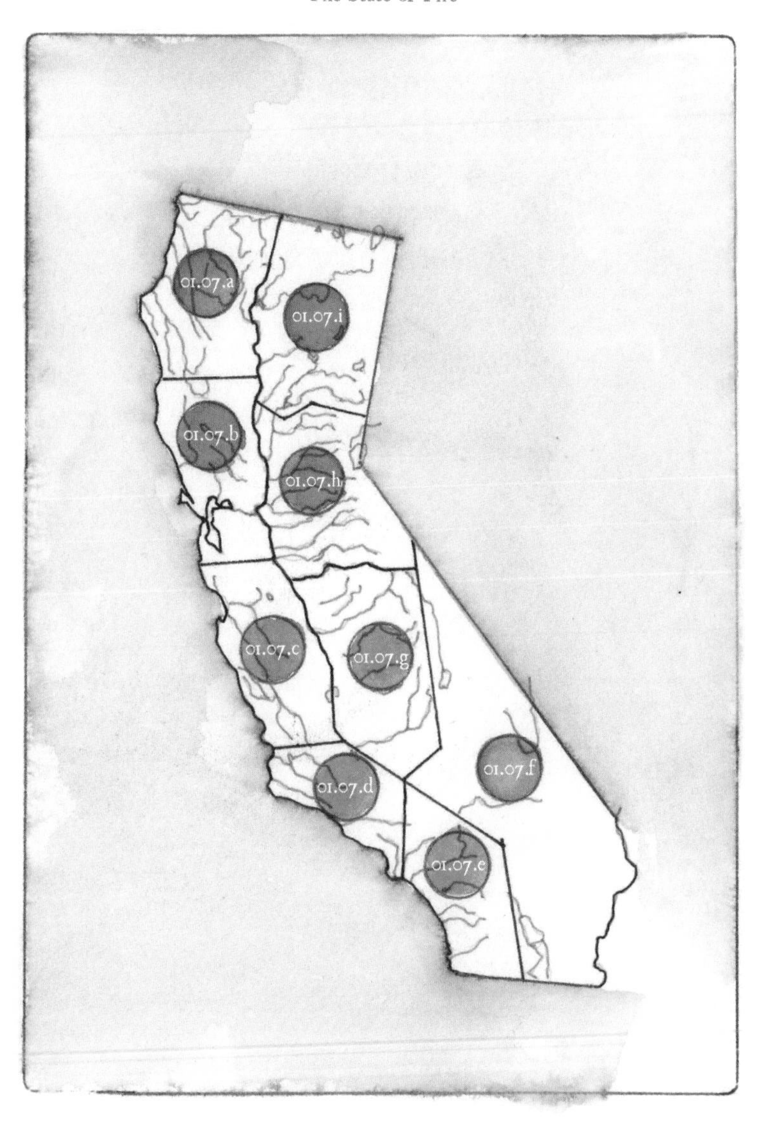
01.07.a
01.07.i
01.07.b
01.07.h
01.07.c
01.07.g
01.07.f
01.07.d
01.07.e

## 01.07 The Century of Reset and Reckoning

### *California fire in the twenty-first century*

What follows are nine maps that show the ranges of many major fires that have occurred since 2000. The maps divide the state into nine geographic regions for formatting convenience; they do not reflect the bioregional divisions detailed later in this book. Many thousands of smaller fires were not included, and the fire's size is not necessarily representative of the fire's significance. For example, in terms of loss of human life and property, sixteen of the twenty most destructive wildfires took place between 2003 and 2023, and most of those fires don't crack the top twenty in terms of overall fire size.[1]

*Some journal notes and commentary are included.*

### *Functional definitions of different kinds of significant fire*[2]

* **Wildfire:** a fire set by nonanthropogenic sources, or escaped fire set by a human source
* **Wildland fire:** a fire that does not impact any settled area; in forest management, wildland is a spatial concept defining an area where the only human infrastructure is that associated with industry and commerce, such as road, rails, power lines, and very few structures[3]
* **Complex fire, or fire complex:** multiple fires over a variety of terrain
* **Fire mosaic:** varying fire intensities and severities across a particular landscape and the habitat types it contains[4]

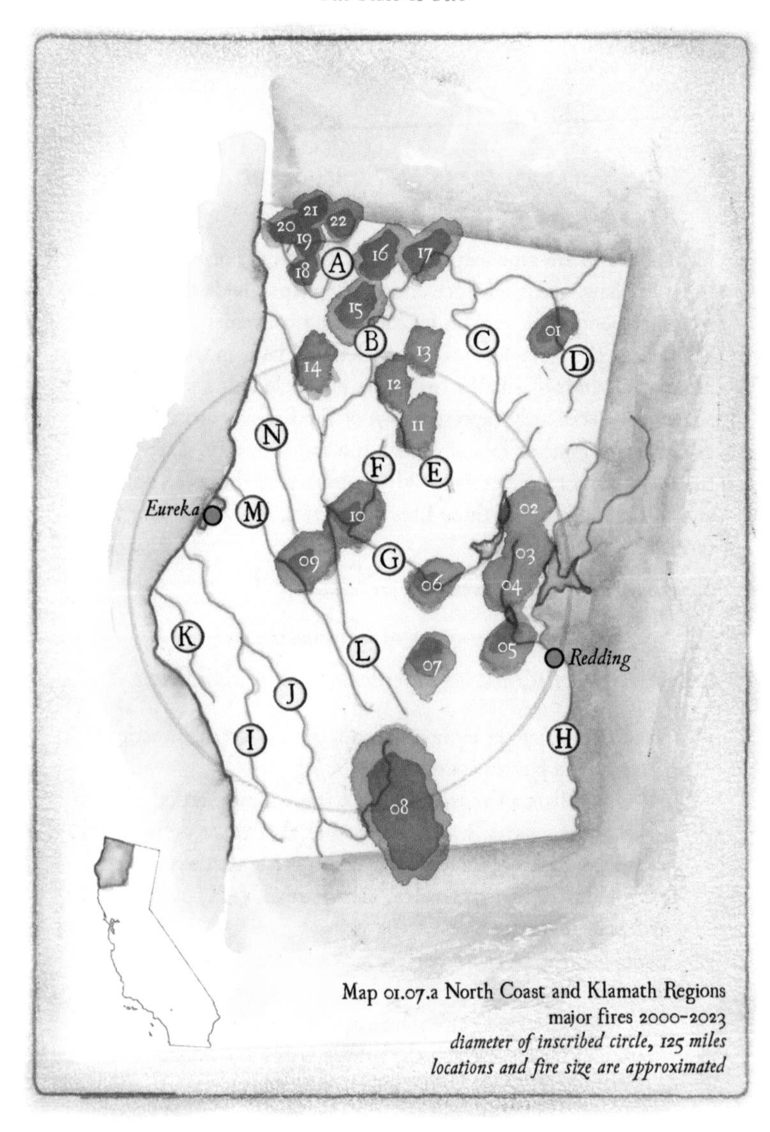


Map 01.07.a North Coast and Klamath Regions
major fires 2000–2023
*diameter of inscribed circle, 125 miles*
*locations and fire size are approximated*

## Map 01.07.a North Coast and Klamath Regions

01. McKinney Fire (2022)—60.1K acres; Shasta River watershed; Siskiyou County
02. Delta Fire (2018)—63.3K acres; Sacramento River watershed; Shasta County
03. Hirz Fire (2018)—46.2K acres; Sacramento River watershed; Shasta County
04. Carr Fire (2018)—229.6K acres; Clear Creek watershed; Shasta County
05. Zogg Fire (2020)—56.3K acres; Cottonwood Creek watershed; Shasta County
06. Monument Fire (2021)—223.1K acres; Trinity River watershed; Trinity County
07. McFarland Fire (2021)—122.6K acres; South Fork Trinity River watershed; Trinity, Shasta, and Tehama Counties
08. August Complex (2020)—1,032.6K acres; Eel River and Sacramento River watershed; Glenn, Lake, Mendocino, Tehama, Trinity, and Shasta Counties
09. Mad River Complex (2015)—73.1K acres; Mad River watershed; Trinity County
10. River Complex (2015)—77.1K acres (and the Fork Complex; 36.5K acres); Trinity River watershed; Shasta County
11. Red Salmon Complex (2020)—144.7K acres; Salmon River watershed; Humboldt, Trinity, and Siskiyou Counties
12. McCash Fire (2021)—94.9K acres; Klamath River watershed; Siskiyou County
13. Iron Alps Complex (2008)—105.8K acres; Trinity River watershed; Trinity County
14. SRF Complex (2023)—50.1K acres; Trinity River watershed; Humboldt and Trinity Counties
15. Klamath Theater Complex (2008)—192K acres; Klamath River watershed; Del Norte and Siskiyou Counties

16. River Complex (2021)—199.3K acres; Klamath River watershed; Siskiyou County
17. Happy Camp Complex (2014)—134.1K acres; Klamath River watershed; Siskiyou County
18. Gasquet Fire (2015)—30.4K acres; Smith River watershed; Del Norte County
19. Eclipse Complex (2017)—78.7K acres; Smith River watershed; Del Norte County
20. Gap Fire (2017)—33.9K acres; Smith River watershed; Del Norte County
21. Smith River Complex (2023)—95.1K acres; Smith River watershed; Del Norte County
22. Slater/Devil Fire (2020)—166.1K acres; Smith River watershed; Del Norte County

A. Smith River
B. Klamath River
C. Scott River
D. Shasta River
E. Salmon River
F. New River
G. Trinity River
H. Sacramento River
I. South Fork Eel River
J. Eel River
K. Mattole River
L. South Fork Trinity River
M. Mad River
N. Redwood Creek

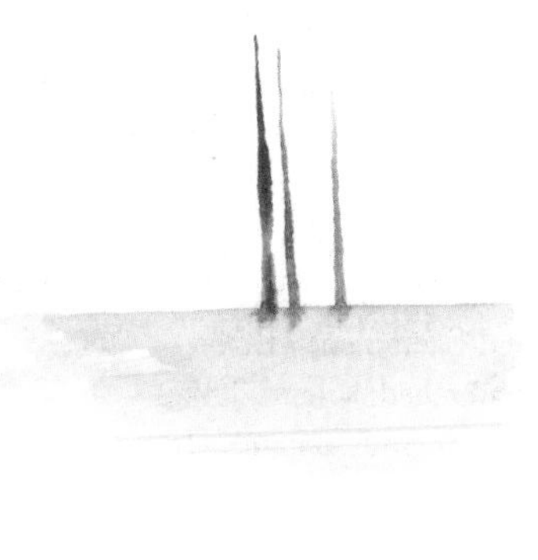

Pyrocumulus

**The Gigafire:** From August 18 to November 12 of 2020, in the biggest fire year in California's history as a state, several regional fires eventually merged to become the first in a category of fire never recorded in North America: the gigafire that was the August Complex Fire. The burning of one million acres began with a lightning storm that lasted a few days, sparking nearly forty backcountry fires in very hot and dry conditions. Buffeted by Diablo wind patterns, the fire jumped in size over twenty-four hours on September 11 from nearly 400,000 acres to almost 700,000.

this one wiped the slate clean. this one changed minds. the year 2017 was rough and dry. So was 2018. the winter of 19/20 was wet and was followed by an extremely hot, windy, and stormy summer. The new growth on parched ground and the climate whiplash together conspired to bring four million acres of burn to the state. How many more four-million-acre years of extensively severe burns are we going to get? What will California be after such a barrage of severe fire everywhere all the time? there can be no more business as usual. We know the answers. turns out that it has been as yet impossible to walk the talk.

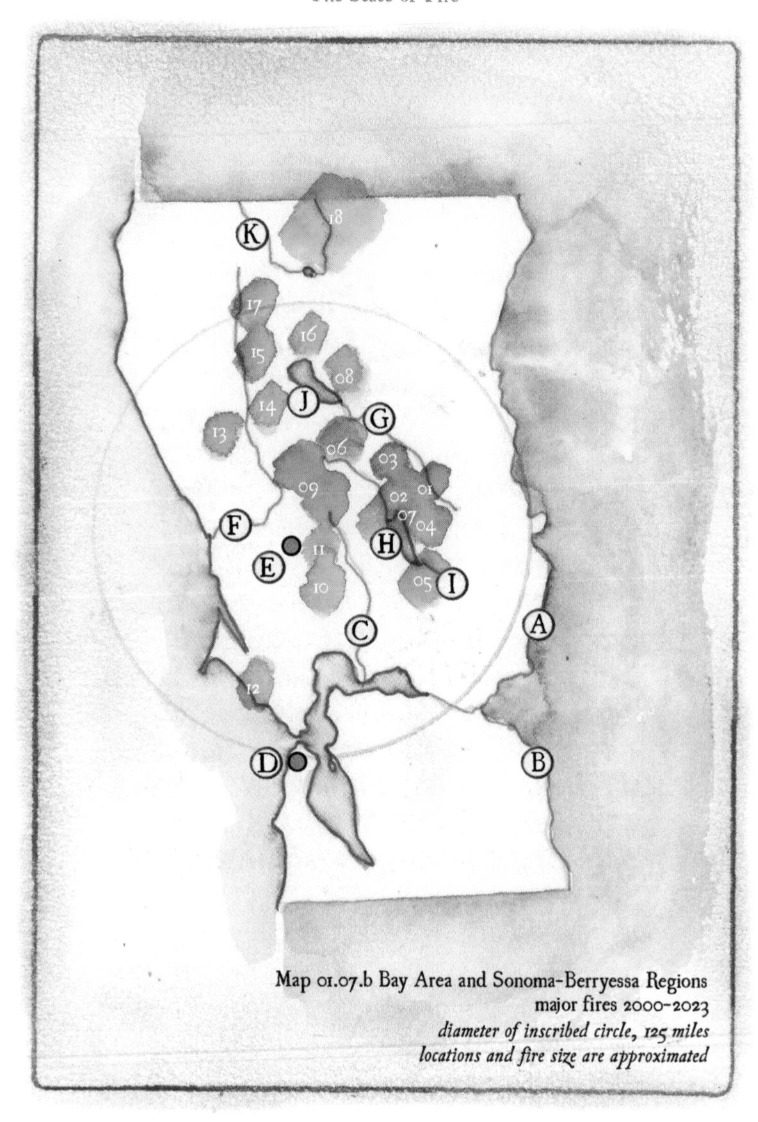


Map 01.07.b Bay Area and Sonoma-Berryessa Regions
major fires 2000–2023
*diameter of inscribed circle, 125 miles*
*locations and fire size are approximated*

## Map 01.07.b Bay Area and Sonoma-Berryessa Regions

01. County Fire (2018)—90.1K acres; Putah Creek watershed; Napa and Yolo Counties
02. Jerusalem Fire (2015)—25.1K acres; Putah Creek watershed; Napa and Lake Counties
03. Rocky Fire (2015)—69.4K acres; Clear Lake watershed; Lake County
04. Hennesey Fire (2020)—305.6K acres; Putah Creek watershed; Napa, Yolo, and Solano Counties
05. Atlas Fire (2017)—51.6K acres; Putah Creek watershed; Napa County
06. Valley Fire (2015)—76.1K acres; Putah Creek watershed; Lake County
07. Rumsey Fire (2004)—39.1K acres; Putah Creek watershed; Napa County
08. Walker Fire (2008)—14.5K acres; Clear Lake watershed; Lake County
09. Kincade Fire (2019)—77.7K acres; Russian River watershed; Sonoma County
10. Tubbs Fire (2017)—36.7K acres; Russian River watershed; Sonoma County
11. Glass Fire (2020)—43.5K acres; Russian River and Napa River watersheds; Sonoma and Napa Counties
12. Nuns Fire (2017)—55.8K acres; Sonoma Creek watershed; Sonoma County
13. Woodward Fire (2020)—4.9K acres; coastal watershed; Marin County
14. Walbridge Fire (2020)—55.2K acres; Russian River watershed; Sonoma County
15. Pocket Fire (2017)—17.4K acres; Russian River watershed; Sonoma County
16. Mendocino Complex; River Fire/Ranch Fire (2018)—459.1K acres; Clear Lake watershed; Mendocino, Lake, Colusa, and Glenn Counties

17. Redwood Valley Fire (2017)—36.5K acres; Russian River watershed; Mendocino County
18. August Complex (2020)—1,032.6K acres; Eel River and Sacramento River watershed; Glenn, Lake, Mendocino, Tehama, Trinity, and Shasta Counties

A. Sacramento River
B. San Joaquin River
C. Napa River
D. San Francisco
E. Santa Rosa
F. Russian River
G. Cache Creek
H. Lake Berryessa
I. Putah Creek
J. Clear Lake
K. Eel River

**The Northern California Firestorm of 2017:** In early October of 2017, the Tubbs Fire, one of eight fires simultaneously burning across the wine-country counties of Northern California, ripped through the city of Santa Rosa, destroying almost six thousand homes and killing twenty-two people.

the city was cut like a burning razor in just a few minutes. Flashover is a process of instantaneous vaporization. Flashover was everywhere. Some homes didn't burn—they vaporized. there was nothing left. No sink. no stove. Just concrete pads and toxic gray dust.

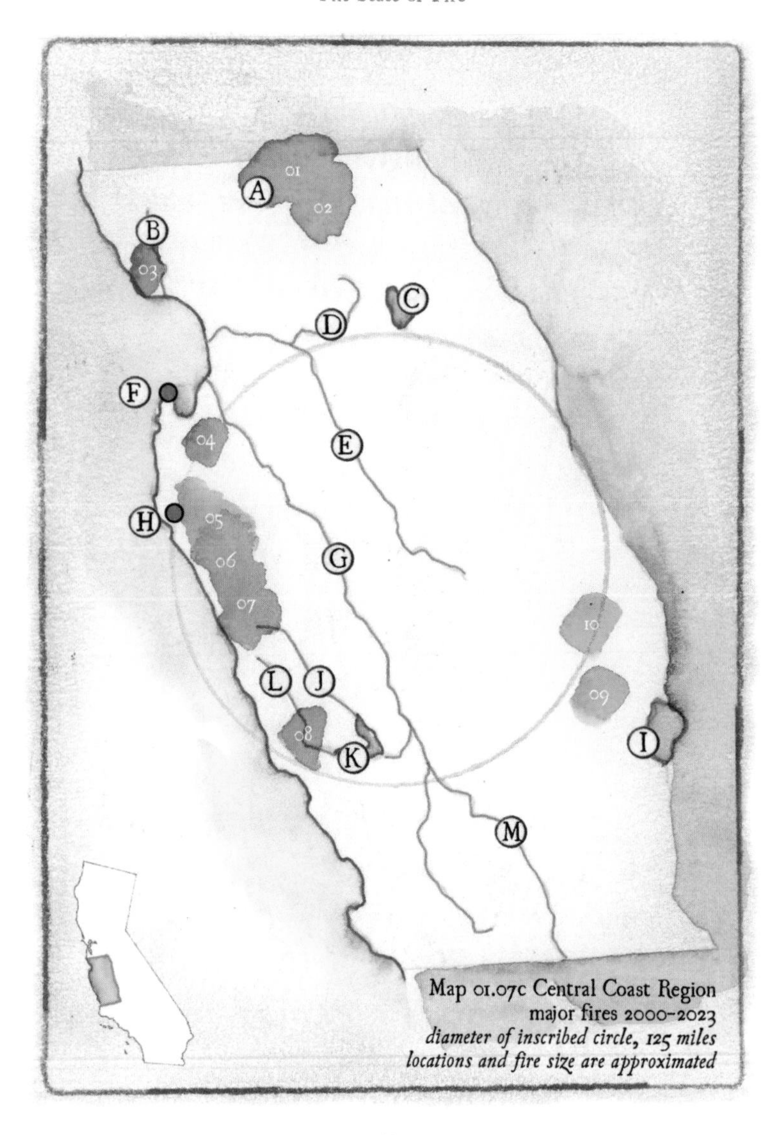


Map 01.07c Central Coast Region
major fires 2000–2023
*diameter of inscribed circle, 125 miles*
*locations and fire size are approximated*

## Map 01.07.c Central Coast Region

01. SCU Complex (2020)—396.6K acres; Coyote Creek and Orestimba Creek watersheds; Santa Clara, Contra Costa, Alameda, Stanislaus, and San Joaquin Counties
02. Lick Fire (2007)—47.8K acres; Coyote Creek watershed; Santa Clara County
03. CZU Complex (2020)—86.5K acres; coastal watersheds; Santa Cruz County
04. River Fire (2020)—48K acres; Salinas River watershed; Monterey County
05. Soberanes Fire (2016)—132.1K acres; Carmel River watershed; Monterey County
06. Basin Complex (2008)—162.8K acres; coastal watersheds; Monterey County
07. Dolan Fire (2020)—124.9K acres; coastal watersheds; Monterey County
08. Chimney Fire (2016)—46.3K acres; Nacimiento River watershed; Monterey and San Luis Obispo Counties
09. Garza Fire (2017)—48.9K acres; Warthan Creek watershed; Monterey and San Benito Counties
10. Mineral Fire (2020)—29.6K acres; San Benito River watershed; Monterey and San Benito Counties

A. Henry W. Coe State Park
B. San Lorenzo River
C. San Luis Reservoir
D. Pajaro River
E. San Benito River
F. Monterey
G. Salinas River
H. Big Sur
I. Tulare Lake
J. San Antonio River
K. Lake Nacimiento
L. Nacimiento River
M. Estrella River

**Fires in the Santa Lucia Mountains:** Scorching ridgelines above Big Sur, the lightning-ignited Basin Complex was the first of three megafires to afflict coastal Monterey County between 2008 and 2020. Eight years after the Basin Complex, which as of 2023 was the twentieth-largest wildfire in California history, the Soberanes Fire raged through rural Monterey County for nearly four months. Soberanes became at the time the most expensive wildfire-fighting effort in history, and at that moment, the Chimney Fire was burning in the Nacimiento River watershed to the south. Even in an area with a long ecological history of recurrent fire, the Soberanes, stoked by an extremely hot summer and started by an unattended campfire, was too much for some vegetation, and took as one of its victims the world's largest Pacific madrone (*Arbutus menziesii*). The arson-sparked 2020 Dolan Fire in the Silver Peak region killed ten critically endangered condors. It was the last of a series of conflagrations that spanned twelve years.

Most of these fires took place in and around the Ventana Wilderness. Wilderness is synonymous in the twenty-first century for places left to burn.

Leaving wilderness to burn
amounts to a big experiment:
if fire is allowed, but people -
meaning cultural stewards
or land treatment and
management are not -
will the quality of the
diverse, living place
not fade? Seems like
a terrible gamble.

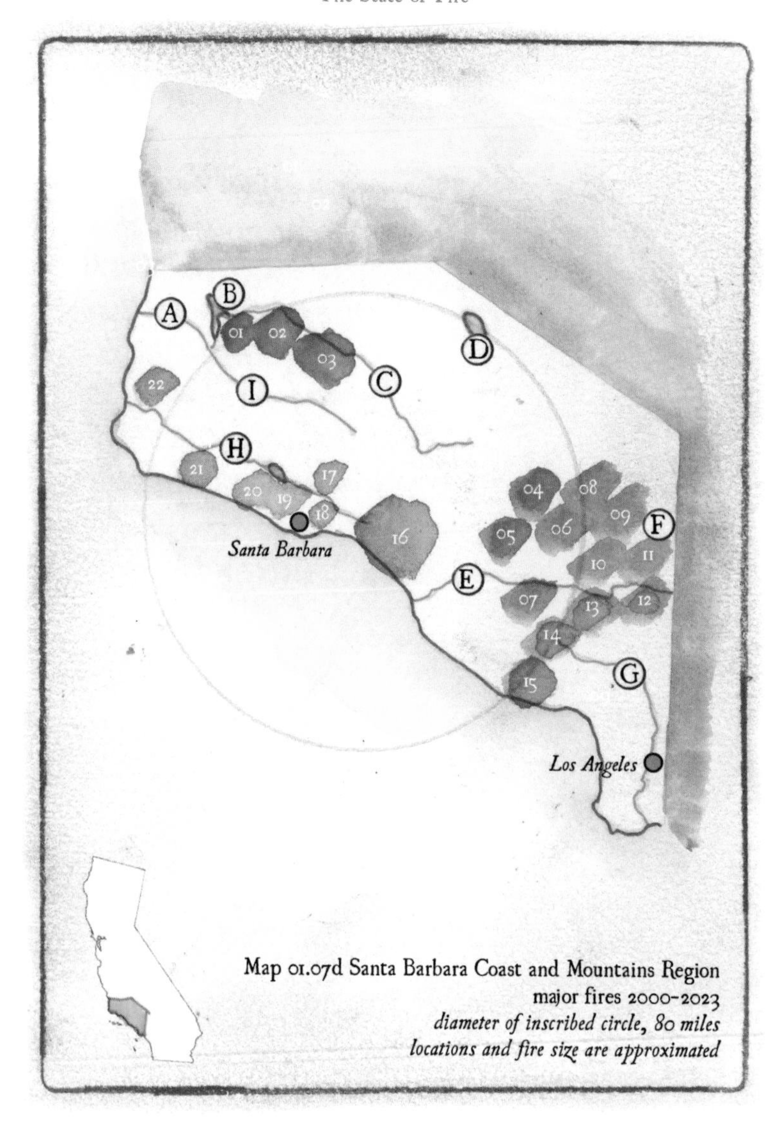


Map 01.07d Santa Barbara Coast and Mountains Region
major fires 2000–2023
*diameter of inscribed circle, 80 miles*
*locations and fire size are approximated*

## Map 01.07.d Santa Barbara Coast and Mountains Region

01. Alamo Fire (2017)—28.7K acres; Alamo Creek watershed; Santa Barbara and San Luis Obispo Counties
02. La Brea Fire (2009)—89.5K acres; Sisquoc River watershed; Santa Barbara County
03. Zaca Fire (2007)—240.2K acres; Sisquoc River watershed; Santa Barbara County
04. Day Fire (2006)—162.7K acres; Piru Creek watershed; Ventura and Los Angeles Counties
05. Piru Fire (2003)—64.0K acres; Sespe Creek and Santa Clara River watersheds; Ventura County
06. Ranch Fire (2007)—58.4K acres; Piru Creek watershed; Los Angeles County
07. Simi Fire (2003)—108.2K acres; Santa Clara River watershed; Ventura and Los Angeles Counties
08. Lake Fire (2020)—31.1K acres; Castaic Lake watershed; Los Angeles County
09. Powerhouse Fire (2013)—30.1K acres; Castaic Lake watershed; Los Angeles County
10. Buckweed Fire (2007)—38.4K acres; Castaic Lake watershed; Los Angeles County
11. Sand Fire (2016)—41.4K acres; Tujunga Wash watershed; Los Angeles County
12. Creek Fire (2017)—15.7K acres; Tujunga Wash watershed; Los Angeles County
13. Sayre Fire (2008)—11.7K acres; Tujunga Wash watershed; Los Angeles County
14. Topanga Fire (2005)—24.2K acres; Calleguas Creek watershed; Ventura and Los Angeles Counties
15. Woolsey Fire (2018)—96.9K acres; coastal watersheds; Ventura and Los Angeles Counties
16. Thomas Fire (2017)—281.9K acres; Ventura River watershed; Ventura County

17. Rey Fire (2016)—32.6K acres; Santa Ynez River watershed; Santa Barbara County
18. Jesusita Fire (2009)—8.7K acres; coastal watersheds; Santa Barbara County
19. Gap Fire (2008)—9.4K acres; coastal watersheds; Santa Barbara County
20. Whittier Fire (2017)—18.4K acres; Santa Ynez River watershed; Santa Barbara County
21. Gaviota Fire (2004)—7.4K acres; coastal watersheds; Santa Barbara County
22. Canyon Fire (2016)—12.7K acres; Santa Ynez River watershed; Santa Barbara County

A. Santa Maria River
B. Lake Twitchell
C. Cuyama River
D. Buena Vista Lake
E. Santa Clara River
F. Castaic Lake
G. Los Angeles River
H. Santa Ynez River
I. Sisquoc River

**The Sayre Fire:** In just six days of November 2008, the Sayre Fire at the WUI—fueled by hurricane-force Santa Ana winds—destroyed nearly five hundred homes, becoming the most destructive wildfire in Los Angeles history.

Every piece of urban infrastructure at the wildland-urban
interface could use a redesign. Fire should govern the
of architecture and design for everything—how food
is produced, how transportation is negotiated, and
especially how homes are conceived and built

...It felt like game over
It felt like all the forms of nature
were made of vapor and smoke.
It felt like the reality of objecthood in the
landscape was a veil for fire to reveal as
ash. And now there are scars, six years later,
but to the uninitiated, to those who don't know,
it looks like the Thomas fire barely happened
at all. —Ventura county

**Thomas Fire:** The Thomas Fire was powerful enough to generate a pyrocumulus cloud, powerful enough to make its own weather. At its height, the fire traveled at one acre per second and did billions of dollars in damage to the community of Ojai.

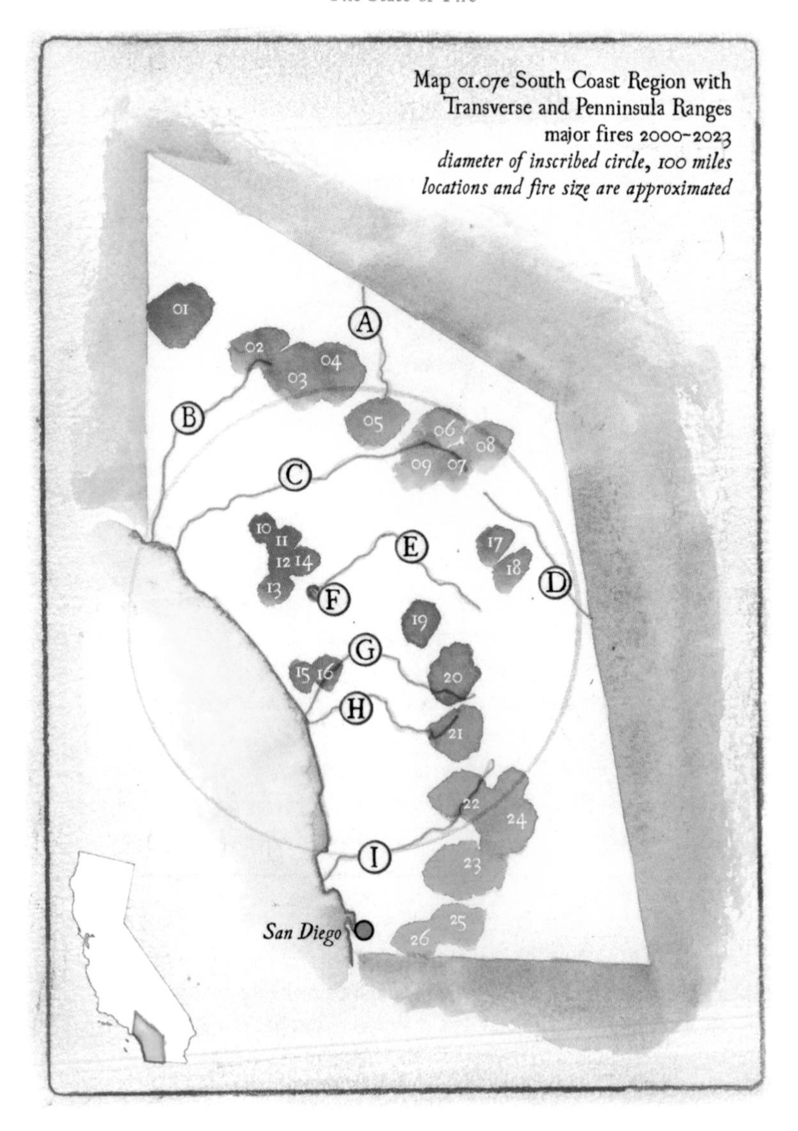
Map 01.07e South Coast Region with
Transverse and Penninsula Ranges
major fires 2000–2023
*diameter of inscribed circle, 100 miles*
*locations and fire size are approximated*
01
02
03
04
05
06
07
08
09
10
11
12
13
14
15
16
17
18
19
20
21
22
23
24
25
26
A
B
C
D
E
F
G
H
I
San Diego

## Map 01.07.e South Coast Region with Transverse and Peninsula Ranges

01. Station Fire (2009)—160.6K acres; Castaic Lake and Bouquet Creek watersheds; Los Angeles County
02. Williams Fire (2002)—38.1K acres; San Gabriel watershed; Los Angeles County
03. Grand Prix Fire (2003)—69.9K acres; Lytle Creek watershed; Los Angeles and San Bernardino Counties
04. Blue Cut Fire (2016)—37.1K acres; Lytle Creek watershed; San Bernardino County
05. Old Fire (2003)—91.2K acres; Santa Ana watershed; San Bernardino County
06. Lake Fire (2015)—31.4K acres; Santa Ana watershed; San Bernardino County
07. Millard Fire (2006)—24.2K acres; Santa Ana watershed; Riverside County
08. Sawtooth Complex (2006)—61.7K acres; Whitewater River watershed; San Bernardino County
09. Apple Fire (2020)—33.4K acres; San Gorgonio River watershed; Riverside and San Bernardino Counties
10. Freeway Complex (2008)—30.3K acres; coastal watersheds; Orange and Riverside Counties
11. Sierra Fire (2006)—10.6K acres; coastal watersheds; Riverside County
12. Santiago Fire (2007)—28.4K acres; coastal watersheds; Orange and Riverside Counties
13. Holy Fire (2018)—23.1K acres; coastal watersheds; Riverside County
14. Cerrito Fire (2004)—16.5K acres; San Jacinto River watershed; Riverside County
15. Ammo Fire (2007)—21.0K acres; Santa Margarita

River watershed; San Diego County

16. Pulgas-Bailone Complex (2014)—26.0K acres; Santa Margarita River watershed; San Diego County
17. Esperanza Fire (2006)—41.2K acres; San Jacinto River watershed; Riverside County
18. Cranston Fire (2018)—13.1K acres; Bautista River watershed; Riverside County
19. Mountain Fire (2003)—10.0K acres; inland creek watersheds; Riverside County
20. Poomacha Fire (2007)—49.4K acres; Escondido Creek watershed; San Diego County
21. Paradise Fire (2003)—56.7K acres; Escondido Creek watershed; San Diego County
22. Witch Fire (2007)—198.9K acres; San Diego River watershed; San Diego County
23. Cedar Fire (2003)—273.2K acres; San Diego River watershed; San Diego County
24. Pines Fire (2002)—61.7K acres; inland creek watersheds; San Diego County
25. Harris Fire (2007)—90.0K acres; Sweetwater River and Otay River watersheds; San Diego County
26. Mine and Otay Fires (2003)—92.3K acres; Otay River watershed; San Diego County

A. Mojave River
B. San Gabriel River
C. Santa Ana River
D. Whitewater River
E. San Jacinto River
F. Lake Elsinore
G. Santa Margarita River
H. San Luis Rey River
I. San Diego River

**The 2003 firestorm:** The year 2003 heralded a new kind of recurrent crown fire in the suburban chaparral, one that would become all too familiar in the years ahead. The Cedar Fire killed fifteen people in its rampage across San Diego County and at one point expanded to engulf almost four thousand acres per hour.

*There is a fundamental disconnect between chaparral fire and urban development. Chaparral needs to expunge itself with hot and fast crown fire, but only twice a century, not every few years. Urban development seems to have an almost metaphysical ignition aura, a radius where fire can't help but find purchase, seemingly analogous to how a lamp spreads light.*

**May 2014 wildfires in San Diego County:** The 2003 firestorm was in November, but the fourteen fires that made up the 2014 wildfires of San Diego County occurred in May. Most of the fires were ignited by human activity, but were stoked by unusual soaring temperatures and hot winds.

Climate anomalies, such as days of 100 degrees in May, will become more and more common. So will fires that arrive in May. This is one reason why the phrase "climate breakdown by way of anthropogenic global warming" is more exact than "climate change," though it may be a mouthful.

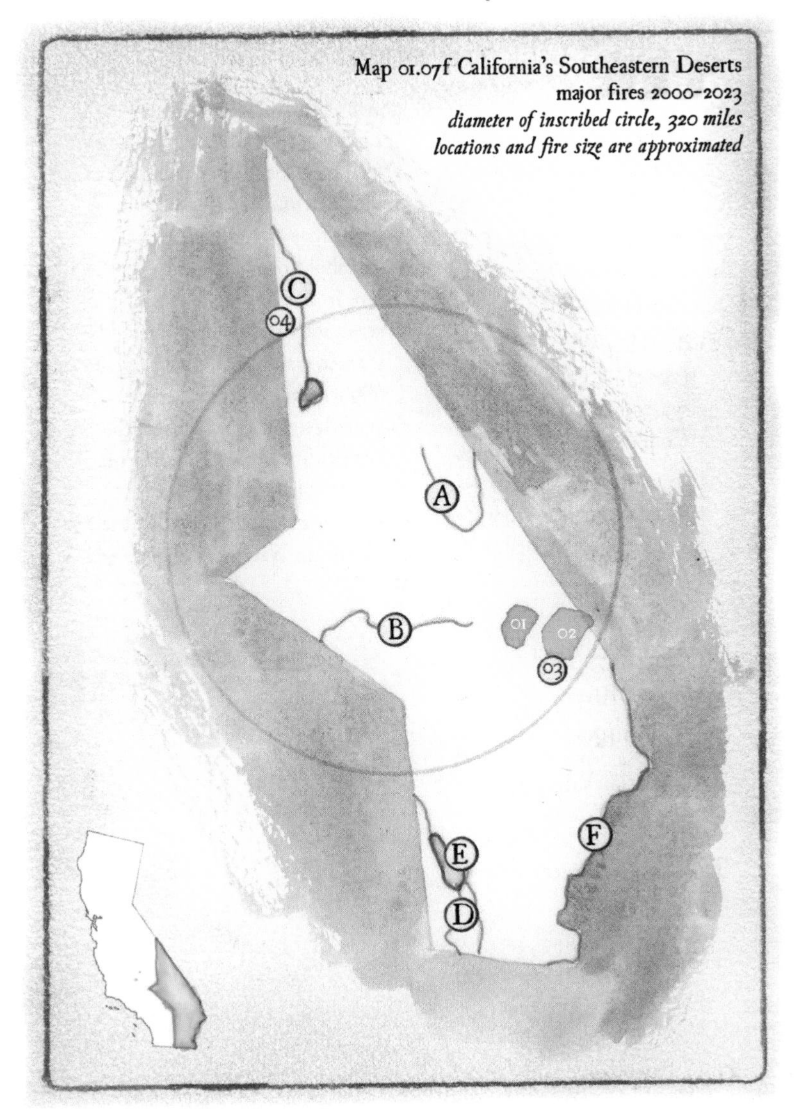
Map 01.07f California's Southeastern Deserts
major fires 2000-2023
*diameter of inscribed circle, 320 miles*
*locations and fire size are approximated*
C
04
A
B
01
02
03
E
D
F

## Map 01.07.f California's Southeastern Deserts

01. Dome Fire (2020)—43.0K acres; desert watersheds; San Bernardino County
02. York Fire (2023)—93.0K acres; desert watersheds; San Bernardino County
03. Hackberry Complex (2005)—70.7K acres; desert watersheds; San Bernardino County
04. Inyo Complex (2007)—35.2K acres; desert watersheds; San Bernardino County

**Dome and York Fires:** Killing well over a million Joshua trees, the lightning-sparked Dome Fire crossed the disturbance threshold of the desert ecosystem in the Mojave National Preserve with furious power, and the ecosystem's adaptive cycle might never recover. Three years later, the York Fire, the largest fire in all of California in 2023, had the same effect when it crashed through Castle Mountain National Monument.

A. Amargosa River

B. Mojave River

C. Owens River

D. Imperial Valley

E. Salton Sea

F. Colorado River

The deserts of California are asked to do too much. Developed and razed for energy production, they are not seen as the delicate and precious systems of endemic life that they are. Perhaps in the interest of protecting what certainly are the unraveling threads of California desert biodiversity, it is time to draft a new designation of federal endangerment that might protect every endemic plant and animal species. Different from a wilderness area designation or what is afforded through the Endangered Species Act, this holistic legislation could bridge the divide between the climate and biodiversity crises, laying the groundwork for new systems of land stewardship and mitigation akin to ancient local practices throughout these arid lands.

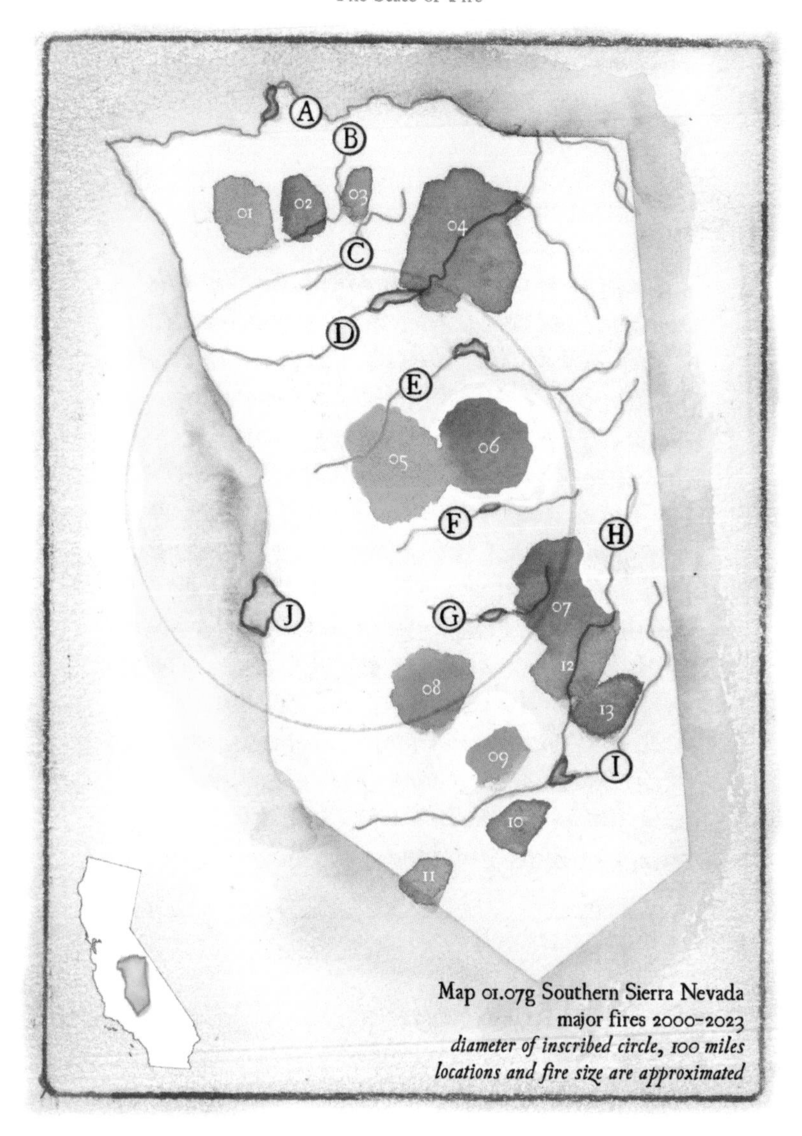


Map 01.07g Southern Sierra Nevada
major fires 2000–2023
*diameter of inscribed circle, 100 miles*
*locations and fire size are approximated*

## Map 01.07.g Southern Sierra Nevada

01. Detwiler Fire (2017)—81.8K acres; Merced River watershed; Mariposa County
02. Ferguson Fire (2018)—96.9K acres; Merced River watershed; Mariposa County
03. Railroad Fire (2012)—12.4K acres; Merced River watershed; Madera County
04. Creek Fire (2020)—379.9K acres; San Joaquin River watershed; Madera and Fresno Counties
05. Rough Fire (2015)—151.6K acres; Kings River watershed; Fresno County
06. KNP Complex Fire (2021)—88.3K acres; Kings River and Kaweah River watersheds; Fresno and Tulare Counties
07. SQF Complex Fire (2020)—174.2K acres; Kern River watershed; Tulare County
08. Windy Fire (2021)—97.5K acres; Kern River watershed; Tulare County
09. Cedar Fire (2010)—29.5K acres; Kern River watershed; Tulare County
10. Erskine Fire (2016)—48.0K acres; Kern River watershed; Kern County
11. Comanche Complex Fire (2011)—29.3K acres; Kern River watershed; Kern County
12. McNally Fire (2002)—150.7K acres; Kern River watershed; Tulare County
13. Manter Fire (2000)—67.3K acres; Kern River watershed; Tulare County

A. Merced River
B. Chowchilla River
C. Fresno River
D. San Joaquin River
E. Kings River
F. Kaweah River
G. Tule River
H. Kern River
I. South Fork Kern River
J. Tulare Lake

**The KNP Complex and the Windy Fire:** After the Castle Fire (SQF Complex Fire) of 2020, which was fueled by an over-crowded, drought-stressed, and beetle-ridden forest, thousands of giant sequoia trees died. A year later, the KNP Complex and the Windy Fire continued the onslaught, lashing many of the same giant sequoia groves, destroying nearly sixteen thousand of the great trees. If giant sequoia, known for their legendary fire resistance, are succumbing to this new kind of supercharged fire, what might that mean for the future of California's forest across the Sierra Nevada?

The beetle-induced tree death followed by the Creek Fire's tree culling seems like an exaggeration of an ancient adaptive succession, a lining up of the pins and then a knocking of them down.

**Creek Fire:** The huge and remote Creek Fire was started by a dry lightning storm in early September 2020 and contributed to California's largest wildfire year since statehood. Fueling the fire was a glut of dead trees, left from the drought-stoked native beetle invasion that lasted from 2012 to 2016 and killed more than 150 million trees in the southern Sierra. Two fire tornadoes were reported in the first days of the firestorm. At one point in the epic story, the National Guard used military helicopters to rescue over two hundred people who were trapped by flames on all sides.

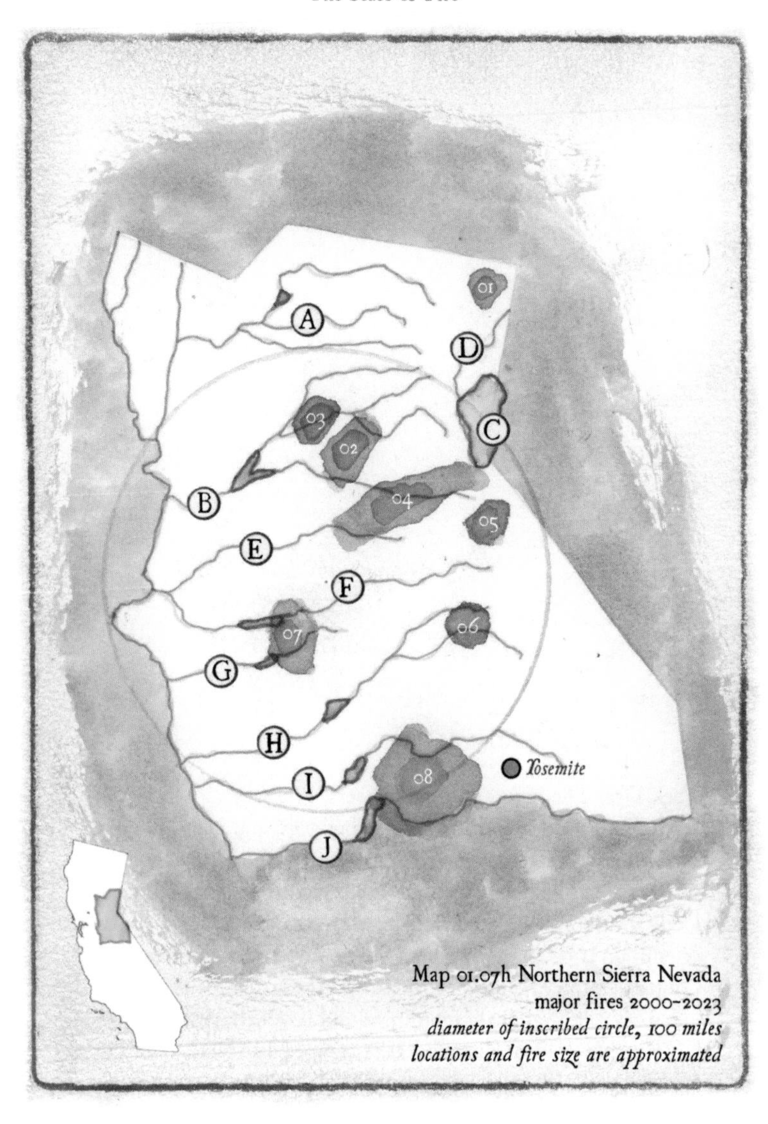


Map 01.07h Northern Sierra Nevada
major fires 2000–2023
*diameter of inscribed circle, 100 miles*
*locations and fire size are approximated*

## Map 01.07.h Northern Sierra Nevada

01. Loyalton Fire (2020)—47.0K acres; Feather River watershed; Sierra County
02. King Fire (2014)—97.0K acres; American River watershed; El Dorado County
03. Mosquito Fire (2022)—76.8K acres; American River watershed; Placer and El Dorado Counties
04. Caldor Fire (2021)—221.8K acres; American River watershed; El Dorado County
05. Tamarack Fire (2021)—68.6K acres; Carson River watershed; Alpine County
06. Donnell Fire (2018)—36.5K acres; Stanislaus River watershed; Calaveras County
07. Butte Fire (2015)—70.9K acres; Mokelumne River and Cosumnes River watersheds; Amador County
08. Rim Fire (2013)—257.3K acres; Tuolumne River watershed; Tuolumne County

A. Yuba River
B. American River
C. Lake Tahoe
D. Truckee River
E. Cosumnes River
F. Mokelumne River
G. Calaveras River
H. Stanislaus River
I. Tuolumne River
J. Merced River

**Rim Fire:** In 2013, the Rim Fire spilled over into Yosemite's National Park with great severity. Ignited by an illegal campfire, the fire spread through a landscape that had suffered from fire exclusion, did not have many big trees, and was overcrowded by both young and aged trees due to past logging. One year after the fire, spotted owl nesting sites in high-severity burn zones achieved the same occupancy rates as they had when they were unburned.

almost five thousand
people were working to
contain this fire at its height.
Witness the good-willed, strong, noble
People determined to contain this force
of the world, this megafire, and all
megafires they are summoned to.

Map 01.07i Northeast Region and Cascade Ranges
major fires 2000–2023
*diameter of inscribed circle, 100 miles*
*locations and fire size are approximated*

## Map 01.07.i Northeast Region and Cascade Ranges

01. July Complex Fire (2020)—83.2K; Lost River watershed; Modoc County
02. W-5 Cold Springs Fire (2020)—84.8K; Great Basin watersheds; Modoc County
03. Gold Fire (2020)—22.6K acres; Pit River watershed; Modoc County
04. Rush Fire (2012)—271.9K; Great Basin watersheds; Lassen County
05. Walker Fire (2019)—54.6K acres; Indian Creek watershed; Plumas County
06. Long Valley Fire (2017)—83.7K acres; Great Basin watersheds; Lassen County
07. Dixie Fire (2021)—963.3K acres; Feather River, Mill Creek, and Butte Creek watersheds; Butte, Plumas, Lassen, Shasta, and Tehama Counties
08. Camp Fire (2018)—153.3K acres; Feather River and Butte Creek watersheds; Plumas County
09. North Complex Fire (2020)—318.9K acres; Feather River watershed; Plumas County

A. Tule Lake
B. Clear Lake
C. Lost River
D. Goose Lake
E. Sacramento River
F. McCloud River
G. Pit River
H. Shasta Lake
I. Eagle Lake
J. Sacramento River
K. Mill Creek
L. Lake Almanor
M. Honey Lake
N. Butte Creek
O. North Fork Feather River/Indian Creek
P. Middle Fork Feather River
Q. Lake Oroville

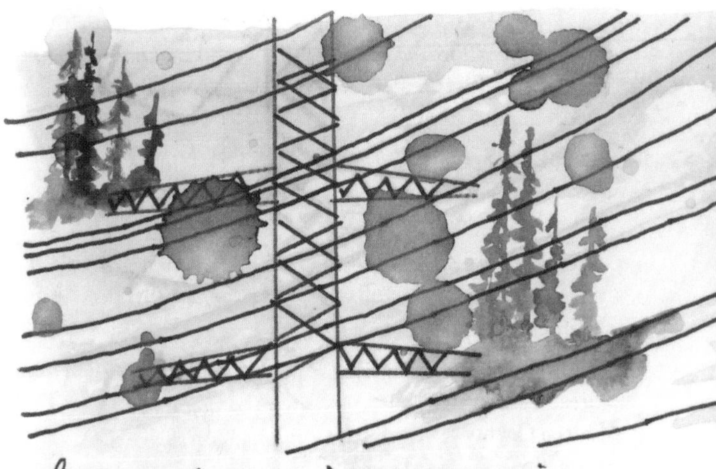

fire waiting, carried through wires
hanging in narrow hallways cut
through fire-prone ecologies.

**Camp Fire:** This fire caused the highest human death toll of any fire in the history of the state of California. Eighty-five people lost their lives as the fire destroyed four mountain towns, including the town of Paradise. In a zone that has seen over a dozen large fires in only twenty years, the highly volatile, logging-altered landscape is always ready to burn. A faulty electric line in the backcountry was enough to get the tragedy under way.

**Dixie Fire:** Close to the destroyed town of Paradise, three years after the Camp Fire, the Dixie Fire began its procession, and it nearly became the state's second record-setting gigafire in as many years. The cause of this colossal fire's rapid spread was not the fierce winds that stoked the Camp Fire but the remoteness and dryness of its spark origin, deep in a ravine, where a tree fell on another PG&E power line.

*Canyon Live in coastal California mosaic habitat*

# PART TWO: FIRE ECOLOGY

## 02.01 Cascading Patterns of Emergence

### *The (new) ecology of fire*

In the twentieth century, when academics and policymakers would talk about fire ecology, they often meant fire *exclusion*. The attitude was that less fire meant more timber for the economy, and it was generally assumed that ecosystems everywhere would thrive if fire did not exist. Although there are aspects to this myth that still exist in the popularized narrative about the destructive specter of wildland fire, over the past few decades there has been a shift in thinking about the value of fire as a regenerative force in California's woodlands, shrublands, and forests.[1] The academic study of fire ecology has become its own respectable discipline, one focused not on the maximization of timber yields. Now divorced from and transforming forestry departments everywhere, fire ecology demands of forestry that it become something other than an agricultural pursuit wherein the economic value of timber is the measure of the forest's total value.[2] Fire ecology examines the dynamic complexity of particular ecosystems in terms of how the structure and configuration of each unique system are adapted to and mediate fire's regular return within varying parameters of intensity and severity.[3] This study dovetails with ecosystem theory, or the study of abiotic and biotic relationships in any given region.[4] These relationships determine what effect fire, as a disturbance event, has on the ecosystem and include climate and weather stressors, botanical associations, vertebrate animal distribution and behavior, fungal propagation, soil, and hydrology (floods, erosion, groundwater, and sedimentation).[5] The new science of fire analyzes how and why fire returns energy to the ecosystem, supporting the ecosystem's functions and services.[6]

Although most of the tenets of fire ecology were established over fifty years ago, recent experiments and observations are reframing fire as a restorative force in California's ecology, one whose relationship to the land is dynamic and symbiotic. What follows are many examples of this push-pull relationship and how it influences the evolution, characterization, and structure of life and fire in California. The term *fire mosaic* is most commonly applied to the description of the effect of different intensities and severities of wildfire on the land, although here it is a shorthand substitute describing mixed-severity fire through various habitat landscapes within a watershed and across different regimes and return intervals.[7] The other term liberally interpreted in this list is informally described as *feral fire*, which is a particularly devastating kind of Anthropocene fire where over an extended area there is little diversity of burn severity, and plant mortality is often 100 percent.[8] Feral fire is a flavor of chaotic fire that may engulf massive land areas that are not adapted to such extreme conflagration and maybe suffer ecotype change because of it. Feral fire might radically alter a woodland, for example, that is unable to recover in time before it becomes a pastureland dominated by invasive grasses.

Wildland fire is a reaction and not a thing. To understand where fire is or will be, you must understand what fire does. It involves itself in a circular set of relationships, both influencing an ecosystem and being influenced by it. Those relationships determine the health of the ecosystem and its ability to recover and regenerate after a disturbance.

The following is a sampling of theoretical fire effects in some California ecosystems:

## ecological engineers / keystone species

An ecological engineer is any organism that, by way of making its own habitat, provides niche resources for other species that inhabit the ecosystem. A keystone species is any organism without which a cavalcade of events occur that will unravel the normal behavior of any ecosystem. The North American beaver (*Castor canadensis*) is both an ecological engineer and a keystone species. This beaver existed in great numbers across the California Floristic Province before the nineteenth century. Its removal greatly diminished a host of associated flora and fauna, altered the local water table (due to the loss of the beaver dam along watercourses), simplified the regional food web, and exposed the landscape to increased aridity and drought vulnerability that exacerbates feral fire to this day.[9] The ability of beaver reintroduced in watercourses to mitigate the most catastrophic effects of feral fire cannot be understated.[10]

## bird populations

Evolving with high-severity fire produces adaptive behaviors that may or may not be obvious. After a high-severity fire, the song sparrow, *Melospiza melodia* (and the other twenty-nine species of sparrow in California), and other migrating songbirds experience an initial drop in population. That drop precedes a population expansion following from increased nest availability in emergent snag forest ecosystems and riparian habitat that might develop following flood events caused by fire.[11] Another bird that may take advantage of fire is the endangered California spotted owl (*Strix occidentalis occidentalis*). The spotted owl benefits from mosaic fire, requiring the regular return of high-severity wildfire for optimized hunting conditions under an open canopy, as well as the shaded security of the mature forest—also called a late-seral forest or sometimes an old-growth forest—where it nests.[12]

## soil fecundity

Mosaic fire's chemical, biological, and physical effects are often a net positive for ecological productivity. Soil is a dynamic medium that combines living and nonliving ingredients. It is true that fire reduces the soil's moisture content by raising its temperature, which can further affect its ability to absorb water; in the short term, these dynamics may decrease nutrient availability. But it is also true that fire volatizes nitrogen deposition from urban pollution, mediating the unwanted nitrogen input and creating a healthier soil environment.[13] The rise in soil temperature also increases the amount of phosphorus available, effectively generating a fertilization effect for seeds that are looking to profit from the fire event.[14]

## water quality

Because mosaic fire reduces plant cover and makes soil harder to penetrate, postfire erosion events are to be expected and can often be catastrophic. Destructive debris flows can change watercourses, transforming the structure of any riparian habitat.[15] In the long term, water temperature may rise due to reduced tree cover. But even though the water's dissolved oxygen content may decrease, the riparian and riverine ecosystems may benefit from increased levels of fertilizing nutrients such as nitrogen, phosphorous, and calcium.[16]

## pathogenic transmission

In some situations, the regular return of mixed-severity fire can reduce invasive fungal and viral pathogens to the ecosystem. In all twenty-five native oak species (*Quercus* spp.) in California (and in tanoak, *Notholithocarpus densiflorus*), fire scars develop as defenses against future fire events. The scars close off the living part of the tree (cambium) from further injury, including biological infiltration, by producing specialized cells (tyloses).[17] Mixed-severity fire may also reduce habitat for pathogenic insect species such as ticks.[18]

## bee and flower diversity

Home to over sixteen hundred bee species, California has perhaps more robust bee diversity than anywhere else in the world.[19] But almost half of California's bee populations are in decline. Fire exclusion is part of the problem.[20] Not only do hundreds of wildflower species depend on postfire habitats to germinate, but plant species tend to produce more flowers per plant than they did before a fire.[21] Coevolving with ever greater numbers of pollinating species such as bees and butterflies, wildflowers have increased their diversity in postfire conditions throughout the Cenozoic.[22] The pollinating services offered by native bees serve to aid the spread of wildflowers that in turn add to fixed nitrogen in the soil, setting up a feedback system that leads to postfire recovery of forest habitat.[23]

## grass evolution

Of the invasive grass species imported to California over the past several hundred years, most evolved on the massive grasslands of the Eurasian steppe, with fire that existed in yearly return intervals.[24] These grasses want to burn and transform California woodlands, forests, and chaparral into the type of environment in which they evolved, and such an environment, it turns out, is good for grazing. Their ecological trajectory is in line with the vision of the colonizing Europeans and Americans. Virulent and persistent, invasive grass species are one of the primary vectors of ecosystem type change, which is the process of destroying endemic adaptive cycles of succession and establishing a novel, fire-prone, simplified ecosystem of grasses, sedges, forbs, and other annuals.[25]

## apex predators

The more complex an ecosystem's food web, where any given prey species might be subject to multiple predators, the more structurally resilient the ecosystem is to disturbance. The predatory force is a profound equalizer in the development of the homeostatic conditions that maintain populations. Often the greatest consequences of this dynamic rest in the behavior of the local apex predator. An apex predator is the so-called top of the food chain and is not prey to any other species in its ecosystem, save for possible human predators. The behavior of the apex predator, as an individual and as a population, has cascading effects throughout the food web. The local extirpation of the apex predator, such as a mountain lion, will transform the local ecology and carries potentially disastrous effects.[26] One such cascading effect with the removal of mountain lions, for example, might be that a local deer population explodes in number, triggering a corresponding explosion in the local deer tick population. The deer population might then begin to overgraze, affecting seasonal erosion patterns on hillsides and in riparian areas, altering local patterns of hydrology and botanical structure, thereby transforming region-wide fire regimes.

Poaceae the colonizing plant family
the grasses

invasive grass species of particular virulence:

Red brome, Bromus madritensis ssp. rubens
Cheatgrass, Bromus tectorum
Yellow starthistle, Centaurea solstitialis
Scotch broom, Cytisus scoparius
Purple veldtgrass, Ehrharta calycina
Medusahead, Elymus caput-medusae
French broom, Genista monspessulana
Scotch broom, Cytisus scoparius
Gorse, Ulex europaeus
Oat, Avena fatua
foxtail, Alopecurus pratensis
Mustard, Brassica nigra

...grasses that are bent
on transforming fire
in California

California
has 1,100 species of non-native
Plants. Nearly 200 of them are
invasive.

Forest succession following high-severity fire: Seral forest progression

01. grass-forb (1-3 years)
02. shrub-seedling (2-10 years)
03. pole-sapling (5-20 years)
04. young (decades)
05. mature (centuries)
06. old-growth (millennia)

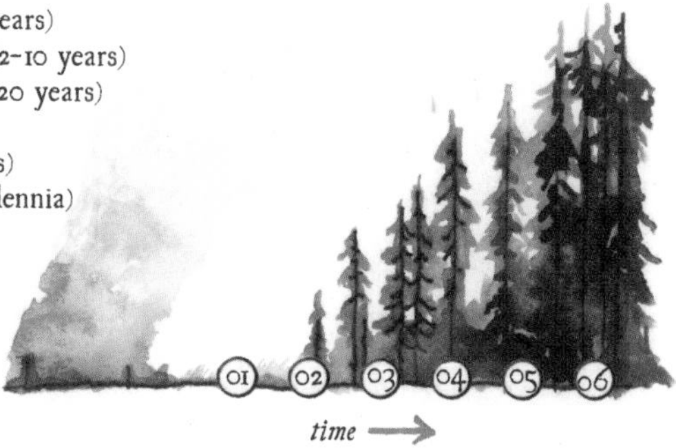

01 02 03 04 05 06

*time* →

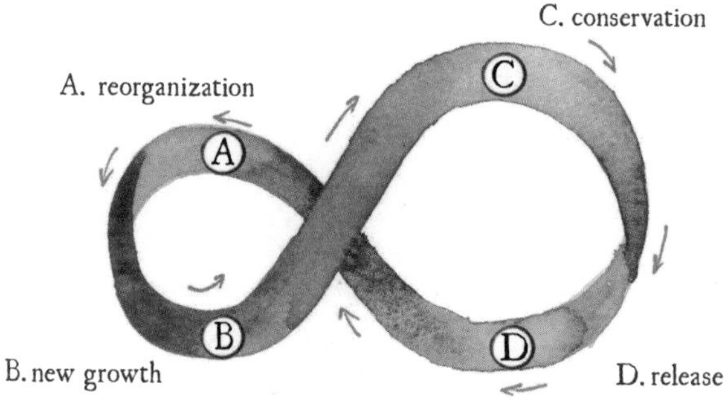

C. conservation

A. reorganization

B. new growth

D. release

The Adaptive Cycle

# 02.02 Succession and Conversion

## *Fire's ecological energy release*

There is an entrenched idea in the Euro-American worldview, spurred by the monotheistic religions of the west and not necessarily shared with the rest of the human species, that imagines the living world (call it Mother Nature) as having something like a mortal destiny. Perhaps it was the ancient theologians, convinced that the world will someday end, who constructed eschatological prophecies about the end times, inspiring the pervasive idea that land fire is intrinsically tragic, an end to all things, and a final point on some imagined ecological timeline. Ecologically, the living world (call it Gaia) doesn't ever end by cataclysm, and the planet's ability to sustain life will only end through effects caused by the evolution of the sun in an inconceivably far future of several billion years.

Wildland fire is determined and managed by the presence of life itself, and from this perspective, fire is something entirely different than an end or a beginning.[1] Ecological wildland fire is part of an adaptive cycle, a wheel that turns, governing the energy of an ecosystem from release to reorganization to growth to conservation, and back again. Fire releases energy, allowing the ecosystem to reorganize its structure, composition, and distribution. That reorganization causes habitat types to grow and mature, which eventually leads to a conservation phase that lasts until fire (or some other disturbance event) returns.[2] This cycle describes the ecological concept of *succession.*

The process of succession describes how any terrestrial botanical ecosystem evolves through time. Ecologists of the early twentieth century determined that the effects of fire within an arboreal or shrub-plant community result in a state of dynamic equilibrium called a fire climax. In a fire-climax community, an ecosystem's health is maintained by being subject to recurring fire of a severity and intensity that does not exceed the ecosystem's resiliency threshold, or its ability to recover postfire, with most of its constituent prefire species present. Homeostasis is the steady state, exemplified by the fire climax, also called late-seral conditions, that may be attained without much loss of total energy. This state is represented by the conservation phase of the adaptive cycle.[3] Nonrecoverable disturbance events, such as massive feral fires or climate change, whether they are recurring or not, lead to type conversion. An example of type conversion might be the demise of a wooded landscape by the annual burning of invasive grass to promote pastureland for livestock. In addition to fire, natural disturbances that spur the adaptive cycle in the ecosystem include flood, windthrow, landslide, insect damage, and extreme weather (such as drought, heat, and high winds). Human activity such as logging, mining, farming, urban development, and air pollution can alter ecological conditions enough to transform local ecosystems and subject them to adaptive recovery or type conversion as dynamically as any natural disturbance.

A fire regime—the pattern that fire follows in an ecosystem—is defined in terms of time, space, and magnitude.[4] More than three hundred fire regimes have been described in California.[5] What follows is a list of attributes that influence them.

Temporal attribute 1: *Fire seasonality.*
When in the calendar year is fire most likely?

* Spring-summer-fall fire season—California's longest fire season, May to November, attributed to fires in the southeastern deserts and low-elevation shrublands
* Summer-fall fire season—California's most common fire season, July through October, attributed to fires in the mixed-conifer coastal forests
* Late summer–short fire season—California's shortest fire season, July through September, attributed to high-elevation lightning fires
* Late summer–fall fire season—California's driest fire season, September through November, attributed to southern chaparral wind-stoked fires

Temporal attribute 2: *Fire return interval.*
How many years are there between fire events?

* Short—1 to 25 years. Examples: Douglas fir and mixed-conifer forests, valley grasslands.
* Medium—25 to 75 years. Examples: chaparral, live oak forests, red fir forests.
* Long—75 to 200 years. Examples: high-elevation conifers, Jeffrey pine, pinyon pine woodlands.

Spatial attribute 1: *Fire size.*
What is the total area burned?

* Small—Less than 25 acres. Example: open pine forest with discontinuous canopy.
* Medium—25 to 2,500 acres. Example: midmontane fir forest with limited burning period.
* Large—Greater than 2,500 acres. Example: grasslands with continuous fuel layer.

Spatial attribute 2: *Fire complexity.*
What are the conditions in which fire takes place?
(These can include transitions from night into day, weather scenarios, seasonal variations, climatic influence, slope and gradient of the landscape, topography, fuel quality and connectivity, and fire history.)

* Low—Homogenous conditions with limited ecological variety. Examples: oak woodlands and chamise chaparral.
* Moderate—Intermediate conditions where burn patches exist across a mosaic of habitat types. Examples: Douglas fir and midmontane ponderosa pine forests.
* High—Burned and unburned mosaic patterns and vegetation mosaics present a large variety of burn conditions. Examples: giant sequoia and mixed-conifer forests.
* Multiple—When multiple complexities occur within a single fire event.

Magnitude attribute 1: *Fire intensity.* How hot is the fire burning, and how high are the flames at the advancing fireline? (Intensity is measured in British thermal units (Btus)—one Btu is the energy it takes to raise the temperature of one pound of water, or a little less than half a liter, by 1° F. For example, a 1,500 kW space heater puts out approximately 85 Btus/second.)

* Low—Less than 100 Btus per meter per second, with flames less than four feet tall; may be managed with hand tools. Examples: annual grasslands and blue oak woodlands.
* Moderate—Between 100 and 500 Btus/meter/second, with flames between four and eight feet tall, in conditions that cannot be managed with hand tools. Examples: giant sequoia forests and coastal sage scrub.
* High—Greater than 500 Btus/meter/second, with flames over eight feet tall. Examples: chaparral and interior live oak in steep canyons.
* Multiple—Different types of fires yielding different conditions, such as the simultaneous burning of a low-intensity surface fire and a high-intensity crown fire. Example: coastal closed-cone pine forests.

Magnitude attribute 2: *Fire severity.* How able is the fire to alter the environment? (Factors include the geomorphology, the hydrology, the habitat spaces, and human property.)

* Low—Most of the vegetation cover remains unchanged, and most individual plants survive. Examples: surface fire in Douglas fir forests and blue oak woodlands.
* Moderate—Mixed intervals of lower and higher severities occur, and most mature plants survive. Examples: Douglas fir forests after fire's long absence, and desert shrublands.

* High—Kills most aboveground plants, with greater than 75% of the dominant overstory vegetation replaced. Examples: quaking aspen and canyon live oak.
* Very high—Causes stand replacement, and all mature plants are killed. Examples: coastal closed-cone forests and ceanothus shrublands.

Magnitude attribute 3: *Fire type.*
How is the fire moving through the vegetation?

* Surface and passive crown fire—Most of the burn is on the ground (surface), with torching of individual trees less than 30%. Example: grasslands.
* Passive and active crown fire—A crown fire is any fire that moves through the canopy of any plant community from plant to plant. A passive crown fire is one supported by surface fire. Examples: mixed-conifer forests where fire exclusion has built up enough ladder fuel to support crown fire.
* Active and independent crown fire—Independent crown fire occurs when fire moves through the canopy without support of surface fire. Chaparral typically burns as an independent crown fire. Examples: conifer forests. Independent crown fires are occurring more recently in conifer forests where they have historically been rare.
* Multiple fire types—Both surface and crown fires are known to occur in the burn regimes of certain ecosystems that specialize in reseeding immediately after fire. Examples: coastal closed-cone forests.

Mean average fire return interval in presettlement fire regimes by ecotype[6]

| *ecotype* | *years* |
|---|---|
| Chapparral | 30 |
| Coastal sage scrub | 20 |
| Oak woodland | 5 |
| Mixed evergreen | 15 |
| Ponderosa pine | 5 |
| Dry mixed conifer | 5 |
| Moist mixed conifer | 5 |
| Red fir | 15 |
| Redwood | 10 |
| Pinyon-juniper | 50 |

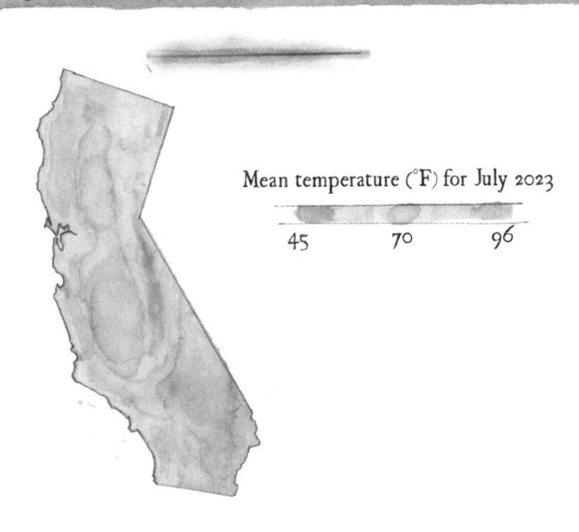

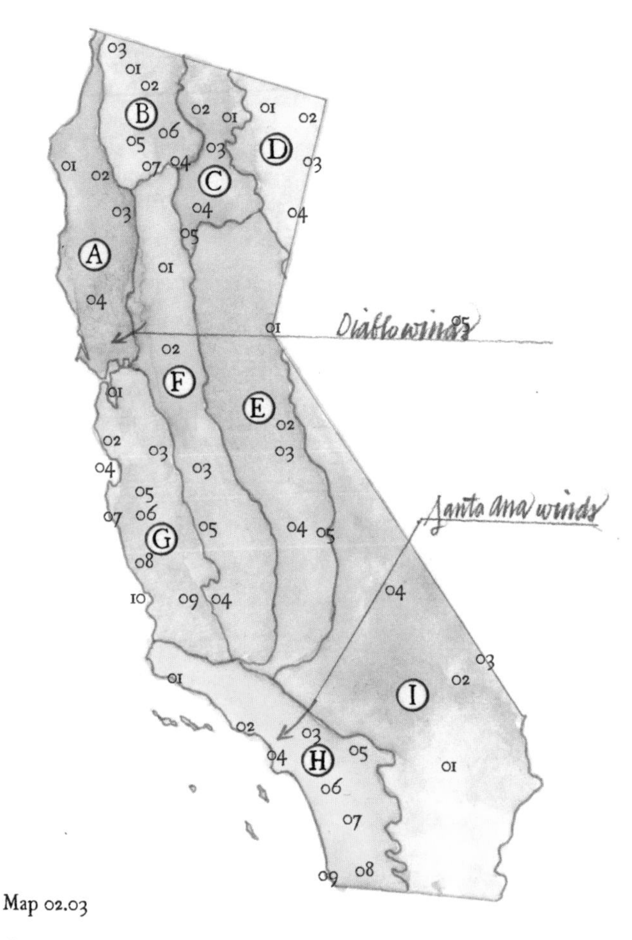


Map 02.03

*Source*: Adapted from S. R. Miles and C. B. Goudey, *Ecological Subregions of California: Section and Subsection Descriptions*, Technical Publication R5-EM-TP-005 (San Francisco: USDA, Forest Service, Pacific Southwest Region, 1997).

## 02.03 Patterns of Vulnerability and Resilience

### *Bioregional fire ecology*

The nine bioregions represented are determined by region-wide relative consistencies of common vegetation patterns, fire regimes, climate types, continental positioning, latitudes, and elevations.

When most people think of wildland fire, they probably think of trees burning. It may be surprising to learn that well over half (64%) of the total burned area from 2000 to 2020 was in nonconifer ecosystems. As its own habitat type, shrublands burned more (38%) than conifer ecosystems (36%). Hardwood ecosystems (oak and riparian habitats) made up 17 percent, and grasslands account for another 10 percent of the total burned area over the first twenty years of this century.[1] More than half of the state's vegetation types by land area (56 million acres of California's total land area of 104 million acres) require the recurrence of fire to maintain their habitat distribution, and often an individual plant's vitality.[2] Studying the bioregions offers a window into why California rivals anywhere else for sheer amount of diversity, whether that diversity is geomorphological or biological.

02.03.a

01. Kings Range
02. South Fork Mountain
03. Yolla Bolly Mountains
04. Mayacamas Mountains

02.03.b

01. Klamath River
02. Marble Mountains
03. Siskiyou Mountains
04. Sacramento River
05. Trinity Alps
06. Trinity Lake
07. Whiskeytown Lake

02.03.c

01. Medicine Mountain
02. Mount Shasta
03. McArthur-Burney Falls Memorial State Park
04. Mount Lassen
05. Paradise

02.03.d

01. Modoc Plateau
02. Warner Mountains
03. Northwestern Basin and Range
04. Great Basin Desert

02.03.e

01. Lake Tahoe
02. Sonora Pass
03. Yosemite National Park
04. Sequoia/Kings Canyon National Park
05. Mount Whitney

02.05.f

01. Sacramento River Valley
02. Sacramento
03. San Joaquin River Valley
04. Arid lands of the western valley
05. Tulare Lake

02.03.g

01. San Francisco Bay
02. Santa Cruz Mountains
03. Diablo Range
04. Monterey Bay
05. Gabilan Range
06. Pinnacles National Park
07. Big Sur
08. Santa Lucia Mountains
09. Temblor Range
10. Morro Bay

02.03.h

01. Santa Ynez Mountains
02. Santa Monica Mountains
03. San Gabriel Mountains
04. Los Angeles
05. San Bernardino Mountains
06. Santa Ana Mountains
07. Vallecito Mountains
08. Laguna Mountains
09. San Diego

02.03.i

01. Joshua Tree National Park
02. Mojave National Preserve
03. Castle Mountains National Monument
04. Death Valley National Park

Journals from the California Floristic Province.
Giant sequoia. Sequoiadendron giganteum...
Biggest. Coastal redwood. Sequoia sempervirens...
Tallest. Cousins. Cupressaceae. Coop-ress-ai-see-eye.
Cypress family. Biggest likes snow. Tallest prefers
to make its own weather with a hundred acres
per tree of flat needles that work like a car
radiator to keep cool and damp. Touch the redwood
and touch 250 million years of genetic legacy.
Touch the profundity of what it takes to survive
the end of the world... many times. Breathe. Two
thousand years in one body. Might have seen
fifty big, big wildfires. Cut it down... it's a crime
of waste, but it doesn't matter. Where there
was one trunk, there will now be six. The plan
is to live for two thousand years, and there is
very little that will prevent that destiny from
being carried out. Fall down? Doesn't matter.
Burn up? Doesn't matter. The little trees die,
but once mature, Big Red is here to stay. Forms
change and the trees persists. Foot-thick bark
insulating and full of fire retardants. Shoots tannin
into the ground too. Keeps the smaller trees out.
Makes more room for the shallow roots and denies

any shade-tolerant ladder fuel from forming.
Chemical warriors. They have to be. There is no room
for any agent whose work is decomposition. No fungus,
No beetle. Antibiotics keep you alive. Be taller
than any flame. Keep the canopy higher than
all the other trees. Let the fire mar the heartwood.
Scar tissue is fire hardened. Tempered. Watch fire
after fires roll through, followed by landslide
after landslide. Doesn't matter. Roots can form
fifty feet up the trunk from buds grown in the
tree body hundreds of years ago. Biggest and
tallest trees have the smallest cones of any conifer.
Fire calls for the cones to open and the seeds to
drift in the wind to find purchase on ground now
shocked by fire fertilizers. One day the big tree
will fall, but it won't be from fire; it'll be from
wind. And then, still chock-full of antimicrobial
agents, its body will rest on the floor for another
few centuries. Plan is to be an ecosystem. A one-tree
forest of sorts. Bathe in fire. Stay clean and strong.

Bay laurel riparian habitat in California

## 02.03.a North Coast and North Coast Ranges Bioregion

Lightning strikes/square mile/year: 2.15[3]

Six-foot pine. Fifty years old. Mendocino's Bolander pygmy forest. Ocean staircase. Small old tree doesn't want fire. Not like the tall and dark forests where fire cleans. Fire would swallow it, as it does all the small trees, and burn the salty soil chemistry. what kind of nightmares does the forest have when it smells smoke only a few miles away?

From sea level to 8,100 feet, the Mediterranean climate of this bioregion experiences a definitive rainy season between October and April, when it receives between 20 and 118 inches of rain—90 percent of its annual precipitation.[4] Summer fog is common throughout the bioregion, and winter freezing events at most elevations are rare. Because lightning ignitions of wildland fire are few and far between, most precontact fire in the region was set by the highly concentrated Indigenous peoples who live throughout the bioregion.[5] Archaeological evidence puts the first people here in significant numbers over eight thousand years ago, millennia before the peak of redwood forest concentrations about five and a half thousand years ago.[6] Dominant native plants in northern coastal scrub and coastal prairies zones—including coyote brush (*Baccharis pilularis*), California oat grass (*Danthonia californica*), and purple needle grass (*Nassella pulchra*, basionym *Stipa pulchra*)—are all facultative sprouters (see 02.05) and are thus representative of populations able to endure short return intervals of high-intensity fire.[7] Northern chaparral shrublands are adapted to high-severity, stand-replacing fires every hundred years or so. Although the seeds of some shrub species are fire-enhanced, others are fire-sensitive, assuring that chaparral germination can take place with or without fire.[8]

Bishop pine (*Pinus muricata*). During the Vision Fire (1995, 12.4K acres, coastal watersheds, Marin County)—which was a high-severity, stand-replacing, wildland fire—the bishop pine forest expanded into the coastal prairie, converting scrub communities to forest and expanding the pine's range by 85 percent.[9] Bishop pine relies on fire to open its serotinous, closed cones, which enables access to the seed. These pines also require fire's effect to open the forest canopy, allowing the seedlings access to light.[10]

Coast redwood (*Sequoia sempervirens*). The tallest tree in the world. Before fire exclusion became Forest Service policy in 1905, nearly sixty years of intensive logging of the old-growth redwood forests were accompanied by slash burning of the new growth in a futile attempt to overcome the persistence of the redwood to survive through asexual postdisturbance regrowth and reproduction.[11] Today, approximately 4 percent of old-growth redwood habitat remains, and because of successive logging treatment, so-called second-growth redwood forests may be even more rare.[12]

Lions and bears. The highest densities of both mountain lions (*Puma concolor*) and American black bears (*Ursus americana*) occur in the north coastal ranges.

## 02.03.b Klamath Mountains Bioregion

Lightning strikes/square mile/year: 9.28

*Mesopredators hunt tiny prey in the shade*
*of the world's most diverse conifer community.*
*Sky-island mountain peaks sit draped*
*with speciated flowers of a thousand-thousand varieties.*
*Rare and ancient tree species hold hands,*
*gasping in the rising temperatures.*

The topography of this bioregion is incised by many deep river valleys, and holds the watershed for California's mightiest coastal rivers, the Klamath and the Trinity. From the bottom of those valleys at 100' to the top of Mount Eddy at 9,025', the Klamath Mountains are Gary Snyder's "mountains and rivers without end."[13] High temperatures, strong winds, and low humidity conspire to bring regular backcountry fires to this remote bioregion. These mountains, due in part to the isolated nature of their fragmented peaks, are one of the most diverse places in the world for conifer species.[14] Characteristic of mixed-severity fire regimes, white fir (*Abies concolor*)–dominated forests of the Siskiyous have a fire return interval of 15–40 years, and so do inland Port Orford cedar (*Chamaecyparis lawsoniana*) forests of the Klamath Mountains.[15] Adapted to low-severity, recurrent, Indigenous fire, Douglas fir (*Pseudotsuga menziesii*) is the Klamath Mountains' most fire-resistant tree species; it weathers fire with deep roots, thick bark, and short needles.[16] One of the region's other fire-resistant conifers is incense cedar (*Calocedrus decurrens*), which can survive severe crown fires thanks to its thick bark and high crown.[17] (There are eighteen conifer species in this bioregion, a world record.)

**Humboldt marten** (*Martes caurina humboldtensis*). A species in consideration for protection under the Endangered Species Act, the marten loves to hunt in the postfire snag forest and shows no signs of fire-caused population decline.[18]

## 02.03.c Southern Cascades Bioregion

Lightning strikes/square mile/year: 13.63

I know lightning fire.
I don't know fire from the heart of Gaia.
I don't know the secrets of those bratty volcanoes
that turn the ground into glass knives
and choke salmon with ashy fingers.

The southern end of a chain of volcanic peaks that extends to British Columbia, the Southern Cascades are, at no more than half a million years old, relatively young.[19] Between the Klamath Mountains and the Modoc Plateau are two landforms dominated by two volcanoes: Mount Shasta (14,179') to the north and Mount Lassen (10,457') to the south. The climate of the region is defined by a cresting axis that runs north to south, separating the more temperate west side from the more arid east side. The southwestern foothills of the bioregion include gray pine (*Pinus sabiniana*), California black oak (*Quercus kelloggii*), and California juniper (*Juniperus californica*). All three species are adapted to low-to-moderate fire return intervals of about twenty years or so; unlogged land burned five times in the last one hundred years and currently has tree populations that average the same diameter as they did in the 1930s.[20] Midmontane conifer forests dominate both the east side and the west side. They generally yield to sagebrush (*Artemisia tridentata*), grass, and shrublands in the north, and to oak, cypress, juniper grass, and shrublands in the south. It is in this bioregion that the high-severity Camp Fire destroyed the town of Paradise in 2018. Due to long decades of fire exclusion, repeated timber harvesting, and overgrazing livestock, a fire in this area was predicted as probable twenty years before it occurred.[21]

The Sierra Nevada red fox (*Vulpes vulpes necator*, state-listed threatened). One of only two existing populations of Sierra Nevada red fox lives in Lassen National Park. Because of the large range that the fox inhabits, fire doesn't affect its numbers as much as other stressors, such as habitat loss, predation by humans, and the reduction of snowpack by global warming.[22] Other regional mesopredators, such as coyote, badger, raccoon, and fisher, are not affected by fire in similar ways.

## 02.03.d Northeastern Plateaus Bioregion

Lightning strikes/square mile/year: 16.30

*Fire in California's northernmost deserts*
*where the plants are the color of smoke*
*and they all die with fire,*
*but they smell so good when they burn.*

In the rain shadow of the western mountains, this arid bioregion is characterized by flat, basalt plains populated by sparse vegetation across the Pit River watershed. The Modoc Plateau is a great plain of shrublands dominated by sagebrush (*Artemisia* ssp.) and dotted with Western juniper (*Juniperus occidentalis*) woodlands and quaking aspen (*Populus tremuloides*) riparian habitat. The Great Basin Desert pours over from Nevada; the piece within this bioregion is called the Mono section because of its ecological continuance with Mono Lake in eastern California. This section of the bioregion is shrub desert and steppe with pockets of single-leaf pinyon pine (*Pinus monophylla*) and Sierra juniper (*Juniperus grandis*). All types of sagebrush except for resprouting silver sagebrush (*Artemisia cana* ssp. *bolanderi*) die from fire, and postfire reestablishment is dependent on the presence of undamaged seeds.[23] Juniper trees less than fifty years old are susceptible to death by fire, but they become more resilient as they age. Presettlement fire return intervals are estimated at 13–25 years.[24] But due largely to the tenacious invasion of cheat grass (*Bromus tectorum*), the five-year period from 2010 to 2014 saw the total number of acres burned increase to at least fourteen times that of any decade since the gold rush.[25]

Greater sage-grouse (*Centrocercus urophasianus*). Eats nothing but sagebrush seeds all winter long. Since the invasion of exotic grass and the uptick of fire, along with an increase in agriculture and development, habitat has diminished for this potentially endangered species.

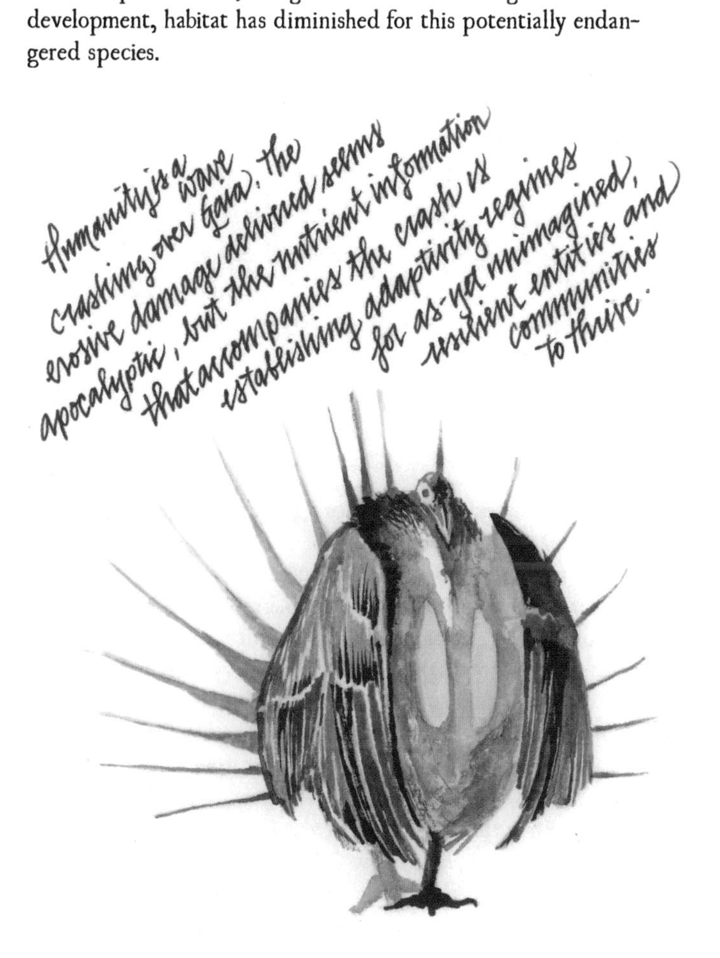

## 02.03.e Sierra Nevada Bioregion

**Lightning strikes/square mile/year: 14.20**

Ten-million-year-old watercourses,
one-million-year-old glacial valleys,
and the trees draw on a legacy
several hundred times older than all of that.
The only soil in this place of granite
is the tree's long-dead ancestors.
Only with some of lichen ally could the
original seed
have pried nutrients from the bare stone.

Topping out at 14,505 feet and with a length of 435 miles, the Sierra Nevada is North America's highest and longest contiguous mountain range. From foothill shrubland to subalpine forest, the many different ecological zones of the Sierra Nevada have been sculpted by fire for millennia, ignited by cultural burn patterns and year-round lightning strikes. Precipitation totals are higher in northern watersheds than southern, and consequently Douglas fir play a more dominant role in the Yuba and Feather River watersheds, which may see sixty inches of precipitation annually. With a cumulative total of 56.4 feet, the snow year of 2023 was the second biggest in the seventy-seven years of recording snowfall totals by the Central Sierra Snow Laboratory, where California's official snowpack record is monitored.[26] This anomalous year of snowfall may be symptomatic of what is called weather whiplash, an indicator of trends in climate breakdown.[27] Throughout the twenty-first century, high severity in mixed-conifer wildfires across the bioregion has increased by an average of 9–27 percent.[28] Region-wide, high-severity fires are increasing in size and thus in postfire erosion potential, and the high severity is also trending toward a reduction of fire complexity and an accompanying reduced capacity for snowpack retention.

The gray wolf (*Canis lupus*) is reclaiming the Sierra Nevada. There are now four wolf packs in the Sierra Nevada: the Lassen, the Beckwourth, the Whaleback, and the Tulare. The presence of these apex predators triggers cascading effects throughout the ecosystem. Famously, when wolves returned to Yellowstone National Park, their presence altered ungulate grazing habits, which in turn altered riparian growth patterns, in turn altering the erosion patterns of rivers, then in turn altering the course of those watercourses. Wolves' effect on grazing routines across the Sierra Nevada may eventually help restore ancient fire regimes in certain habitat types.

At-risk animal species of the Sierra Nevada that require habitat in mature forest stands and are affected negatively by fire exclusion and high-severity fire: Pacific fisher (*Martes pennanti pacifica*), American marten (*Martes americana*), northern goshawk (*Accipiter gentilis*), and California spotted owl (*Strix occidentalis occidentalis*).

## 02.03.f Central Valley Bioregion

Lightning strikes/square mile/year: 2.89

Geologists have called this alluvial plain
a fertile bowl of mud,
This land of flowers and ghost lakes that haunt
the formerly flooded place,
Once saturated with snowmelt
and then with fire:
its yearly breath

With over 80 percent of its total land area developed in modern times, the grass and river lands of California's Great Central Valley are the most altered of any of bioregion.[29] Not only the land has been altered, but also the waterscape: California's two largest watersheds, the Sacramento and San Joaquin river valleys, are subject to complex water supply projects that dictate flows depending on societal needs. Blue oak (*Quercus douglasii*) is an important species in oak savanna, and even though it is threatened by invasive vegetation that competes for water, mature trees survive fire, and blue oak seedling recruitment is unaffected by fire.[30] Presettlement grasslands now exist only as relict populations near protected vernal pools, and they represent 1 percent of their original area.[31] These grasslands once dominated the land area of the valley and were composed of a mix of perennial grasses and forbs that were torched by human fire almost every year.[32] Although fire would have been rare in the estimated one million acres of riparian habitat that existed in the valley two hundred years ago, many common riparian species, such as California sycamore (*Platanus racemosa*) respond vigorously to recurrent low-intensity fire.[33]

North American beaver (*Castor canadensis*) in Dos Rios State Park, at the confluence of the San Joaquin and the Tuolumne Rivers, work at helping with flood control and the reestablishment of riparian habitat in the massively developed Central Valley. With their capability to reinforce local hydrology, beavers could be a valuable piece of the fire puzzle. The state of California is now reestablishing beaver populations at test sites.[34]

## 02.03.g Central Coast Bioregion

Lightning strikes/square mile/year: 2.14

The animals survive fire; they've evolved with it.
They seemed to have also survived a worse threat:
human callousness, waste, and predation.
The numbers of condor, otters, osprey, raptors,
lions, tule elk, whales, and many others are
increasing,
The emergent result of stewardship
born of love and hard work.

Coastal mountain ranges in this bioregion trend from northwest to southeast, and they include the Diablo Range, the longest of California's coastal ranges. Patchy networks of complex, nutrient-poor soils, including serpentine soils in the Santa Lucia range, give rise to many different ecological zones, including those dominated by conifers such as knobcone pine (*Pinus attenuata*), Sargent's cypress (*Cupressus sargentii*), Monterey pine (*Pinus radiata*), and the endemic Santa Lucia fir (*Abies bracteata*). During the rainless season from May to October, marine upwelling causes low stratus clouds in the summer that hang close to California's southernmost grove of coastal redwoods.[35] Although they are separated by only about fifty miles, across Monterey Bay, the Santa Cruz Mountains have burned much less over the past one hundred years than the Santa Lucia Mountains. Rugged and remote terrain, as well as extreme weather, stoked chaparral fires of the Santa Lucias into regular massive conflagrations throughout the twentieth century.[36] With average temperatures along the coast expected to increase dramatically in the coming decades, so too are wildfires in this bioregion expected to get bigger and come more often as the fire season swells, although precipitation and land use patterns may mediate this effect.[37]

Tule elk (*Cervus canadensis nannodes*). Nearly extinct in the twentieth century, California's native tule elk now number about six thousand individuals and graze the ranges of the central coast in herds of up to fifty.[38]

## 02.03.h South Coast Bioregion

Lightning strikes/square mile/year: 7.37

The fire races up the canyon and torches
the big-cone Douglas fir overlooking the city,
and the heavy black pillow of smoke
over the coast is composed of vaporized homes.
Wait for the big rains next year when the mountain
will liquefy and race to the sea.

The east–west–running Transverse Mountains form the northern border of this bioregion and are separated by broad, heavily populated valleys. The north–south–running Peninsular Ranges mark the eastern border of the region and divide the coast from the desert. This bioregion totals less than 10 percent of California's land area, but accounts for over half of the state's human population.[39] Highly altered grasslands and sage scrub habitats dominate lower elevations, yielding to chaparral and oak savanna at middle elevations, with pine forest in montane habitats. The tallest peak, at 11,503 feet, is Mount San Gorgonio at the northeasternmost corner of this bioregion. The entire bioregion is prone to recurrent fire in short return intervals. The intervals are getting shorter, and fires are getting smaller, due in large part to increased development and management.[40] Isolated and dense groves of closed-cone cypress species—including Tecate cypress (*Hesperocyparis forbesii*) and Cuyamaca cypress (*Hesperocyparis stephensonii*)—exist in the Peninsular Ranges and are fire-obligate seeders that require long intervals between crown fires to propagate.[41]

Torrey pine (*Pinus torreyana*), one of the rarest pine trees in the world, exists in only two bluff-top populations: one on Santa Rosa Island and one in San Diego. It clings to the coast partly due to a fire tolerance so low that even low-intensity fire will kill any tree with a trunk of less than a 21-inch diameter.[42]

## 02.03.i Southeastern Deserts Bioregion

Lightning strikes/square mile/year: 19.81

The big grass trees, the yucca trees,
arms waving, can't handle these burns,
and those tortoises that didn't cook in their shells
are scarred and can't live in the gray mess.
The Mojave always changes into something else.
Will it still be the Mojave in fifty years?

The Mojave, Sonoran, and Great Basin Floristic Provinces, on the east side of the Sierra crest, burn differently than the temperate bioregions to the west. With high concentrations of yearly lightning strikes and highly altered regimes due to invasive grasses, many ecosystems in this bioregion are suffering due to increased fire severity. In the Great Basin, invasive grasses have reduced fire return intervals of 30–100 years down to an average of 5 years.[43] For a variety of cultural and food services, the Kumeyaay of the Anza-Borrego Desert and Coachella Valley (the Sonoran Desert, also called the Colorado Desert or the Low Desert) would regularly burn the grassy habitat around freshwater springs that supported California fan palm (*Washingtonia filifera*).[44] Desert conservation challenges abound region wide as increasing nighttime temperatures and growing fire size and severity threaten the iconic Joshua tree (*Yucca brevifolia*) forests of the Mojave's high desert.[45]

Desert tortoise (*Gopherus agassizii*). Because of the desert tortoise's wide range and its downward-trending population, conservation efforts in the desert are regularly centered around the tortoise and the connectivity of its habitat.[46]

*Let the tortoise speak. the story to be told is old and so valuable as to be priceless.*

## Primary ethological vectors determined by fire

Fire attributes: (S) spatial—size and complexity, (T) temporal—seasonality and frequency, (M) magnitude—severity and type. Fire attributes primarily influence three distinct qualities of animal behavior in any affected population: reproductive and rearing success (how, where, and when offspring may be had); migration, colonization, and exploitation (how, where, and when animals move); and eviction and displacement (how, where, and when animals may be forcibly replacted by fire).

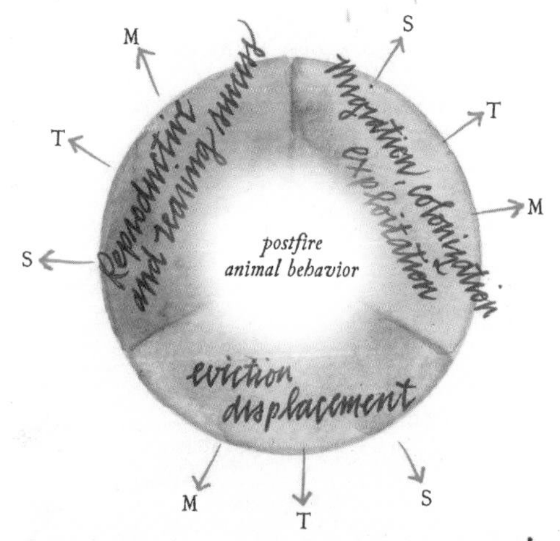


**Maximizing fitness opportunities long-term patterns of opportunities and obstacles**

*Source:* Adapted from M. H. Huff and J. K. Smith, "Fire Effects on Animal Communities," in *Wildland Fauna in Ecosystems*, General Technical Report RMRS-GTR-42, ed. J. K. Smith et al., 1:35–42 (Ogden, UT: USDA, Forest Service, Rocky Mountain Research Station, 2000).

## 02.04 Abundance and Variation

### *How fires affect animals*

Since the advent of three major events—the Last Glacial Maximum, the early Holocene extinction of so many megafaunal species, and the coming of anthropogenic fire regimes—the configuration of animal populations and the fecundity of their habitat have been primarily governed by fire on the land. Mixed-severity fire physically alters habitats across the landscape, presenting greater complexity and novel access to nutrients and other ecological elements. These alterations in the ecosystem influence animals' physiology. Fire's regular or irregular return (time), its trajectory through the ecosystem (space), and its impact on habitat (magnitude) all influence the behavioral and biological timing of many animal characteristics, including their physical displacement, reproductive timing and gestation, the rearing of offspring, and the migration of populations.

The forest fire seen in Walt Disney's *Bambi* (1942), which shows all the animals running in panic from a human-caused fire and is soundtracked with sinister music, did more to influence popular conceptions of animal behavior during a fire than any other piece of twentieth-century American storytelling. The scene suggests that escape is the only recourse for the animals of the forest. But most smaller animals (snakes, lizards, rodents, rabbits, toads, and ground birds) hide underground, insulating themselves from heat, flame, and smoke, and are quickly out and about after the fire has passed.[1] Larger animals, such as bear and deer, have a relatively low mortality rate directly because of fire unless their passage is prevented by a fence or other obstruction.[2] The evidence suggests that for the most part, birds are not frightened by fire and can be found foraging soon after the fire moves through.[3]

While there is no evidence that even the most severe fire event has ever caused the extinction of any species, the stress that fire can inflict on a local population may, in the near future, do so.[4] Post-fire recovery of animal populations is based on changes in plant configuration and distribution.[5] In the twenty-first century, low populations of particular species, alterations to fire regimes, and habitat fragmentation may be exacerbating factors that could be disastrous in the face of a major disturbance event such as a fire.

The snag forest, also called the complex early seral forest, is the habitat that emerges after high-severity fire (or drought, or beetle infestation) moves through. Where black spindles of burned conifer trees are often all that is left, the fire-fertilized soil quickly pushes up green shoots through the gray ash. Countless niche resources become available to a host of enterprising, so-called edge species. The biodiversity metrics of a snag forest are often higher than any other seral forest types.[6] Research suggests that more than half of all California bird species, nearly all migratory mammals, and over 90 percent of butterfly and moth species can take advantage of snag forest.[7] This truth subverts the common intuition that what remains after a high-severity fire is a dead, destroyed piece of land that holds little opportunity for life. In a postfire habitat, populations of edge species (woodpeckers and songbirds, for example, also called open-site species) rise dramatically, and populations of species that rely on understory habitats diminish. Over time, though, successional forces replace those niche resources favored by edge species, and conditions again favor species dependent on understory and old growth.

### Mice and rats

Of the ten different species of native kangaroo rat (*Dipodomys* sp.), wood rat (*Neotoma* sp.), and mouse (*Chaetodipus* sp., *Peromyscus* sp., and *Perognathus* sp.) studied after a high-severity fire in Southern California's coastal sage scrub and in southern chaparral, six of the species' populations survived intact with minimal losses. Nine species were present again with recovered numbers within a year. All ten species exceeded prefire population numbers within three years. This behavior is attributed to a wider array of available fire-following forage and increased shelter opportunities.[8]

### Black-backed woodpecker (*Picoides arcticus*)

The black-backed woodpecker has been a resident of California's charred forests for the past three million years.[9] This small species of woodpecker uses an arsenal of adaptations for life in the snag forest: its body shape and size, uniquely configured skull, and three-toed feet that allow it to strike deeper than other species into burned trees and find tens of thousands of beetle larvae every day. California is the southernmost extent of this boreal bird's North American range, and it is estimated that there are only a few hundred mated pairs of black-backed woodpeckers alive today in the Sierra Nevada because of the lack of the suitable postfire habitat.[10] These birds require habitat created by high-severity fire in old forests. Compared to the historical baseline, perhaps only one-third of such habitat exists today.[11] The primary drivers of this decline are fire suppression and postfire snag logging.

California spotted owl (*Strix occidentalis occidentalis*)

The spotted owl stitches together the fire-mosaiced landscape. There are two major populations of California spotted owls. In 2023, the US Fish and Wildlife Service proposed that the first, the coastal and southern population, be protected under the Endangered Species Act. The other population, the Sierra Nevada population, is listed as a threatened species. Although spotted owls require high canopy cover and old, wide trees for nesting, they prefer to hunt in snag forests more than in any other forest type.[12] A logged postfire snag forest will result in a declined spotted owl population, but fire itself, regardless of magnitude, doesn't seem to affect local spotted owl populations.[13]

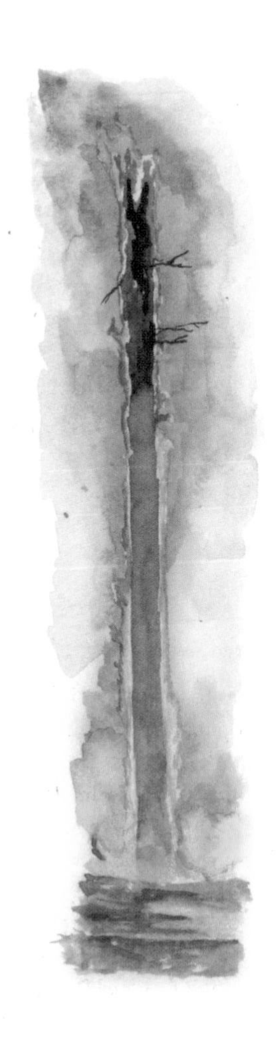

# 02.05 Dependency and Enhancement

## *How fire affects plants*

As fire influences plant growth, so too do plant growth, behavior, and community structure influence fire behavior. Seasonal phenology (plant behavior through the year), the fire return interval (length and regularity of fire frequency in the local ecology), and fire variability (intensity, severity, and spatial complexity) are a few of the primary factors that inform the relationship of plants and fire.

During the 1995 Vision Fire at Point Reyes National Seashore, 3,570 acres of a monotypic forest of bishop pine (*Pinus muricata*) burned with great intensity, producing a complete stand replacement. This local population of pyrophytic bishop pine used the fire's intensity to regenerate a completely new forest. The trees ostensibly sacrificed themselves to the high-severity, low-frequency fire, which allowed the successional seed release from serotinous cones to enable a renewal. On the soil of their arboreal ancestors, the seeds of what would become the new forest found purchase in the fire-shocked, nutrient-rich soil.[1]

The following list of common plant adaptations to fire is grossly simplified and not meant to suggest that any given plant, population, or species only ever may evolutionarily select from one of the listed options as their strategy for responding to fire. Individual plants may exhibit different responses just as different populations might, even within the same species. Chamise, for example, described here as a facultative seeder/sprouter, may be a fire-enhanced sprouter at higher elevations and a fire-neutral sprouter at lower elevations.[2]

Fire-neutral—plants for which sprouting recruitment, the successful emergence of new plants, and/or seed release and seedling establishment occur at similar rates with or without normalized fire regimes

Example: Coast live oak (*Quercus agrifolia*) with its waxy leaves and thick bark, can survive most low-severity fires and is considered fire-neutral in its adaptive capacity to withstand fire events.[3]

Fire-inhibited—plants that do not respond well to disturbance by fire and not only experience individual plant death but often suffer reduced territory and a decrease of population

Example: Most species of fir (*Abies* spp.) are killed by fire and do not have high seedling recruitment following a fire. Firs, especially white fir (*Abies concolor*), colonize fire-prone, montane forests when normalized fire regimes are prevented.[4]

Fire-sensitive—plants that by morphology or seasonality can only tolerate limited fire

Example: Because wet soils conduct ground heat more readily than dry soils, ponderosa pine (*Pinus ponderosa*) roots are vulnerable to burn, so ponderosa's mortality in the spring is up to 30% higher than in equivalent fires in the fall.[5] Understanding this seasonal sensitivity is crucial when considering prescribed fire policy.

Fire-enhanced—plants that become more productive or may expand territory when subject to a normalized fire regime

Example: California fan palms (*Washingtonia filifera*) do not have vascular tissue in the form of cambium, but rather distribute it throughout the tree stem. Fire can't girdle and kill the fan palm, and it leads to increased palm fruit production and reduced transpiration.[6]

Fire-persistent—plants that regrow quickly after fire

Example: Western juniper (*Juniperus occidentalis*) trees are structured to burn quickly and completely to the ground, so the soil temperature is not raised too high and the plant can quickly resprout.[7]

Fire-resistant—plants that withstand damage from fire

Example: Ponderosa pines self-prune their lower branches so that ladder fuels can't accumulate. This adaptation protects buds in the crown from complete destruction, allowing the pine's needles to regrow quickly.[8] Long needles and flaky bark protect the cambium (the living part of the tree under the bark) and diffuse the heat, which also helps these trees survive fire.[9]

Fire-dependent—plants that require normalized fire regimes to assist in some physiological role related to species proliferation

Example: Many fire-dependent, obligate-seeding conifers such as knobcone pine (*Pinus attenuata*) have serotinous cones. Also called closed cones, serotinous cones can remain on the tree and viable in the canopy for up to fifty years. Only at temperatures greater than 320° F will the cone scales open and release the seeds. Generally, serotinous cones hold up to a hundred seeds, with 80% seed viability.[10] Other pyrophytic (fire-adaptive and fire-dependent), closed-cone species include lodgepole pine (*P. contorta*), bishop pine (*P. muricata*), Monterey pine (*P. radiata*), Coulter pine (*P. coulteri*), giant sequoia (*Sequoiadendron giganteum*), and several cypress species and cypress hybrids (*Hesperocyparis* spp.).

Facultative seeder-sprouters—plants that use buried seed banks to persist following fire and can sprout new growth from buds in the stem, boles, or lignotubers, or from belowground.[11]

Example: Chamise (*Adenostoma fasciculatum*) is a common chaparral shrub plant throughout California. Because it can resprout from the ground (basal resprouting) and can germinate from its buried seed bank, it is often the first plant to appear following severe fire.[12]

Obligate seeders—plants that can reproduce only following fire, whether by germination of buried seed bank, seed release by heat, or colonization of burned habitat via long-distance seed dispersal.

Example: Tobacco brush (*Ceanothus velutinus*) is a fire-enhanced shrub whose seed bank is stimulated by high-intensity fire. Bigpod ceanothus (*Ceanothus megacarpus*) only releases its seeds under the high heat conditions supplied by fire. Thistle (*Cirsium* spp.) and willow (*Salix* spp.) have light, feathery seeds that can travel long distances on the wind.[13]

Obligate sprouters—plants that respond to damage from fire (or weather or grazing) by regenerating through new growth

Example: Mariposa lily (*Calochortus* spp.) and camas (*Camassia* spp.) sprout from surface-penetrating bulbs that are spurred to grow by fire.[14]

# PART THREE: FIRE PRINCIPLES

## 03.01 World on Fire

### *California climate breakdown by the numbers*

The world's climate is transforming in human time. The rate of this transformation is much higher than at any point during the Cenozoic.[1] The only analog in the fossil pollen record is at the end of the Young Dryas stadial (stage) (13,800–11,500 YBP), when global average temperatures increased at a similar rate and a massive spike in California wildfire occurred.[2] Like a cavalcade of erupting supervolcanoes, or like the Chicxulub meteorite that spurred the fifth mass extinction, anthropogenic global warming is supplying the fuel for a kind of fire in the sky. Unlike land fire, atmospheric warming doesn't burn, but it does cause a blanketing effect that slowly cooks the biosphere and, like land fire, initiates cascading reactions that sustain and empower it.

The global systems by which the world evolved to its present state are transforming so quickly that the stress of that transformation is threatening those systems' collapse. Abiotic stressors that alter the behavior of wildland fire include drought stress, precipitation whiplash, increased lighting strikes, rising air temperature, and decreasing water availability.[3] The desiccating effect of drought stress on tree and shrub species compromises their standard life cycles, as well as their ability to adapt to and resist biotic invasion and predation.[4] Wild swings of extreme weather, or climate whiplash—expected to become 25 percent more frequent in the next century—is responsible for wet-season expansion of fuel that then dries in subsequent drought events, priming the landscape for more explosive fire.[5] Although precipitation totals are not expected to drop in the twenty-first century, the snow-to-water ratio is expected to drop between 32 and 79 percent, leaving California in something close to a perpetual snow drought.[6] With every degree Celsius of global warming, there is at least a 12 percent increase in wildland lightning strikes, meaning that the likelihood of ignition will also increase.[7] In the next century, the annual mean temperature of California will rise between 4.2° F and 14.8° F, and the number of extremely hot days is expected to rise by at least 23 percent.[8] With the increase in temperature, there is a corresponding decrease in the amount of water available to plants. This effect, called the climate water deficit, is one of the most important measures of ecological effects on plant life, and this deficit is expected to increase by between 1.6 to 6.3 inches over the next one hundred years, suggesting a massive loss in water availability.[9] Climate modeling suggests that by the year 2085, there could be a 200 percent increase in area burned compared to averages from the past two decades.[10] Not only will hot places get hotter and wet places get wetter, but fiery places will get more fiery.[11]

In California, wildland fire and climate breakdown feed each other. In 2020, California's largest wildfire year in modern history, 4.2 million acres burned and emitted 106.7 million metric tons of carbon dioxide ($CO_2$). In the twenty years from 2000 to 2019, total burn area in California averaged .63 million acres per year, with an average yearly emission rate of 14.29 million metric tons of $CO_2$. In 2020, California emitted a total of 369.2 million

may 2023
NPS-WHIS
phoenix flowers
in the snag
forest, open and
nutrient-rich,
now saturated,
every barren
trunk, scarred
and gray, has
new purpose,
utilized by
an endlessly
diverse avian
community
of song birds,
swallows,
woodpeckers, and
other vernal celebrants -
including mammals -
reptiles, and
invertebrates.
The mind of
the forest
seems to know
no
tragedy.

metric tons of anthropogenic $CO_2$ (excluding natural sources). Thankfully, this number was down from the 460 million metric tons California generated in the year 2000.[12] Still, the amount of anthropogenic carbon emissions that California produces by itself—1 percent of humanity's total carbon input into the atmosphere—is roughly equivalent to the yearly average of atmospheric $CO_2$ produced by all volcanic activity on the planet.[13]

Mature forest ecosystems are currently responsible for sequestering about one-quarter of all human-generated carbon emissions, mainly in the biomass of tree wood and leaves. But because bigger and more severe wildfires will exacerbate global temperatures, forests' baseline ability to sequester carbon is rapidly shifting.[14] At the same time, the emergence of "zombie forests," where heat causes certain tree species to be unable to process $CO_2$ and ostensibly breathe correctly, may transform forests from carbon sinks to carbon sources.[15] Because of a so-called climate mismatch brought on by already altered climate conditions, it is estimated that almost 20 percent of the forests in the Sierra Nevada are unable to access adequate nutrient conditions to support their continuance into the future.[16] These hot, stagnant trees are suffocating. They may already be dead and not know it.

The current chaos in California's arboreal ecosystems is distorting ecological norms of the past. Established patterns of recovery following massive fire events may have been thrown out the window by the Anthropocene's atmospheric crush—at least temporarily. For over a century and a half, the suppression of cultural burning has disrupted fire-prone ecosystems' normal response to fires when they do occur. This understanding of historical and cultural fire demands a recontextualizing of the force humanity exerts on the world. Humanity's planetary consequences extend back before the dawn of human cognition, and they will extend to some future point when the climate legacy left by the global human experiment fades and the world becomes something else. The accumulation of so much global data regarding climate breakdown and what effect that has on California land fire demands the evaluation of what responsibility humanity has to its own future.

climate breakdown by way of anthropogenic global warming makes more of every stress: more and longer heat waves, more and longer drought periods, more lightning strikes, more aridity, increased vapor deficit in the vegetation... But this is not the whole story, and it is not a fate already written. All systems of life exist in adaptive cycles and emergent phenomena that are waiting to be studied and waiting to be learned from.

# 03.02 Fire Is the Hunter

## *Human culpability, fire threat, and fire design*

Fire is not a thing. Fire is not an object. Fire is not inanimate. Once it ignites, fire exists in scales of time and space through patterns of behavior and regime. At its smallest scale, fire exists as flame, and flame is the instantaneous combustion reaction that occurs when its constituent components of oxygen, heat, and fuel are ignited. Fire's behavior on a landscape may unfold over a period of days and is bound by parameters of weather, topography, and the availability of fuel. Fire regimes, how fire exists in calendric cycles on a fire-prone landscape, depend on climate conditions, ignition types and frequencies, and vegetation patterns. Fire is not biologically alive, but it does possess many qualities that give the impression that it exists as a form of life. This form of life can be described as ecological. If fire is ecologically alive, it makes sense that it would seem to possess many attributes reserved for living beings. Fire dances, it dies, it adapts, it sleeps, it even breathes—fire seems compelled toward the consumption of fuel in a manner reminiscent of how plants consume sunlight. Unlike other abiotic ecological agents, such as soil type or meteorology, fire by combustion is uniquely positioned in its need to consume biological material to exist for a brief while before it dies. Fire possesses a mindless will to exist. Fire will not be denied. Fire will not end. Fire will not stop.

As the human footprint grows, wildland fire grows. It keeps pace and matches every advance of humanity with its own. Wildland fire seems to be shadowing humanity's legacy footprint in the Anthropocene. Fire sticks close to human activity like a wolf to its prey. Fire is the hunter. Fire Hazard Severity Zone maps illustrate California's Moderate, High, and Very High fire hazard areas, but it also maps the WUI, and it describes a cartographic history of California's industrial and suburban expansion into the fire-prone backcountry over the past century and a half. The shaded regions where the threat exists resemble a wave breaking as urbanizing landscapes topple over and into unsettled landscape blocks. Fire seems to welcome the rapid expansion into its home territory. In the age of feral fire, the cost of basing this continuing expansion on faulty assumptions could not be more dire. It can no longer be business as usual when developing the backcountry. Fire's omnipresence must be carefully weighed with every development decision.

Imagining the novel design of residential landscapes in the WUI is an economy-transforming opportunity. The fire threat is so universal that a fundamental reassessment of basic municipal designs is warranted. Town design, defensible space, metal-roofed dwellings partially built underground, fire corridors, and an array of actively maintained border defenses begin to detail the list of design considerations that should be taken into account when building communities in the WUI. The cost of implementing such infrastructural design is high, but following a fire catastrophe, the cost of not having done so is surely higher. This design methodology might represent a three-pronged approach of resistance (denying *the hunter* its fuel), cocreation (allowing *the hunter* to find fuel in naturalized regimes), and retreat (realizing that *the hunter* will claim its territory).[1] What would the future look like if fire threat zones were uniquely respected, expertly designed, and faithfully stewarded? What if society's relationship with fire could become less antagonistic and California's pyrogeographic history were embraced? *The hunter* remembers the past, and maybe it is time the humans did too.

Modern society is built on containing fire and using it as a tool. The combustion engine is a tool of fire. The electrical cord is a tool of fire. For the most part, human behavior is a function of humanity's relationship with fire. Humans can be very clumsy creatures, regularly unfocused and unable to reconcile their activities with their awareness of the world around them. Combating the unintentionality involved in ignition seems as great a challenge as any.

01. equipment 27.2%
02. power lines 20.5%
03. arson 20.1%
04. lightning 15.9%
05. debris burning 5.1%
06. vehicle 4.4%
07. camping 3.6%
08. playing 1.9%
09. smoking 1.4%

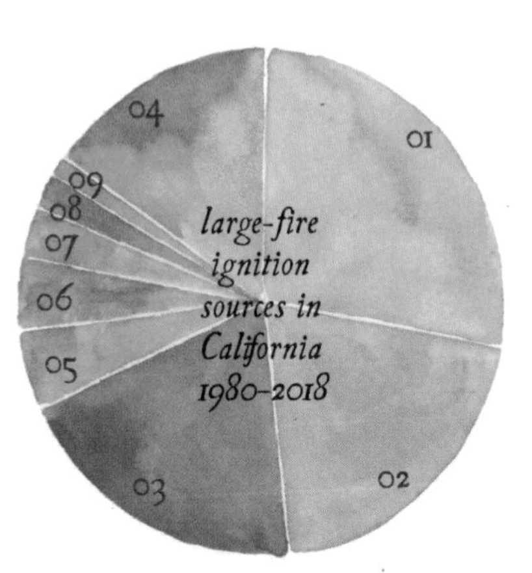


Although arson, or deliberate vandalism, is responsible for a significant percentage of fires, most fires are set by accident. Wherever there is human activity, there will be fire. From 1980 to 2018, 84.1 percent of all fires were human caused, and in that time frame, discarded cigarettes were the cause of more land area burned than was burning for intentional stewardship and land management.[2]

Tree mortality: in 2016, after an anomalously hot, five-year drought, nearly 130 million trees died across one and a half million acres of low montane and midmontane conifer forests in the southern Sierra Nevada.[3] Tree density (primarily Douglas fir, ponderosa pine, and white fir) in this part of the southern Sierra now exceeds six hundred trees/acre. Overcrowded and water stressed, these small, young, and even aged trees are suffering from lack of land management and may never grow old or big. They may never enjoy the habitat conditions that existed only two hundred years ago, when tree density averaged sixty trees/acre.[4] This calamity is not isolated in the southern Sierra and will continue to spread. CAL FIRE reports that of California's total forested 32 million acres, 10 million are infirm forestlands that require treatment—fire, thinning, and harvesting.[5]

# 03.03 A Policy of Prescription

## *Opportunities and challenges in normalizing wildland fire*

Over the next one hundred years, California's climate and its landscape may be determined as much by how the land is adaptively managed as it will be by the industrial mitigation of anthropogenic carbon emissions.[1] The dire need for the complete installation of a comprehensive suite of proactive fire management tools has been clear for decades, but the fulfillment of statewide objectives remains elusive.[2] Part of the challenge is economic, and part of it is ethical. Forest management is expensive and dangerous. Prescriptive fires, for example, may burn out of control, or the smoke may be hazardous to local communities.[3] If the whole suite of management tools were available, it would range from energy infrastructure redesign to the massive escalation of prescribed fire application and would involve private and public industries on a scale not yet fathomed.[4]

One of the great paradoxes of fire is that with more fire, patterns of less fire emerge. Another way to conceive of this is that with more *good fire* comes less *feral fire*. In this case, good fire is recurring, often low-intensity fire that enables ecosystems to remain resilient and robust.[5] Feral fire, by contrast, is high-severity fire that destroys habitats. The idea of bringing fire to the land to protect land from fire is slowly gaining ground. But it will take a long time for this practice to be widely accepted again. Just over forty thousand acres of forestland were treated by CAL FIRE in the 2022/2023 fiscal year. That acreage is approximately 1 percent of forested land that has a high FRI (fire return interval), meaning that it has been way too long since that land has seen fire.[6] Those forty thousand acres also represent 10 percent of CAL FIRE's fuel treatment goals.[7] Just as it took approximately one hundred years for fuel conditions to become as we know them today, it may well take another one hundred years to restore the ecosystems to how they were before.[8]

Current research suggests that although no solution for fuel reduction works as well as fire on the land, mechanical surrogates such as forest thinning and tree mastication—which can be a euphemism for logging—might mediate the effects of high-severity fire.[9] Some commercial timber operators, sometimes working with CAL FIRE, may claim that their work is purposeful thinning and fuel reduction, but they have been accused of removing larger, economically valuable trees while leaving the smaller trees that make the best ladder fuel.[10] Further, some environmentalists argue that by removing trees, you also remove shade and the wind shields that trees may provide, exposing the land to greater fire danger.[11] Regardless, if the intention is fuel reduction for the purpose of increasing resiliency and restoring biodiversity, then thinning only at modest levels (without building new roads) and implementing grazing programs are important tools to moderate devastating crown fires in mixed-conifer forests—if coupled with techniques of surface fuel reduction.[12]

Fuel reduction and management work effectively to reduce megafires and to promote restoration only on systems that are adapted to burn with regular low-intensity fires and not on systems that burn as crown fires, such as chaparral.[13] Megafire transforms, resets, and in some instances restores fuel loads across large areas for great short-term resiliency, but will only continue to do so if future burn management regimes are implemented, sometimes in return intervals as small as three to five years.[14] Management following megafires presents ongoing challenges, as these landscapes often can't reestablish forests because of a lack of seed viability or seed dispersal and often become home to high-density shrubland that presents a continuous fuelbed for the next megafire.[15] The strategy of harvesting dead trees (postfire salvaging of dead trees, as opposed to prefire thinning of live trees) is often presented as a way to mitigate future fires, when in truth it may be more of an economic enterprise, as the results in terms of fire mitigation, restoration, or short-term habitat resources are neutral at best, deleterious at worst.[16]

Current fire practices and land management decisions are affecting fire conditions at least as much as amenable fire weather and fuel-burdened landscapes.[17] For example, increased nature preserve acreage and a growing tolerance of fire's ecological process have led to a generalized policy of letting backcountry fires burn, and decidedly not spending (or risking) the firefighting resources to extinguish them immediately, which contributes to the trend of bigger and bigger fires.[18] In the letting go of the famous 10 a.m. rule—the USFS policy of extinguishing all fires anywhere by ten in the morning the day after they were spotted—early in the twenty-first century, massive wildland fires became inevitable. The success of the subsidence of fire exclusion policy will be measured by the ability to institute prescribed fire regimes and to attain the project goals for treated acreage. Fire's return this century, after its long, actively designed absence, has resulted in the aggregation of conflagration after conflagration, but this result was arguably not unexpected. One of the basic truths about fire that is more apparent now than ever is that California burns because of decisions people make as much as for any other reason.

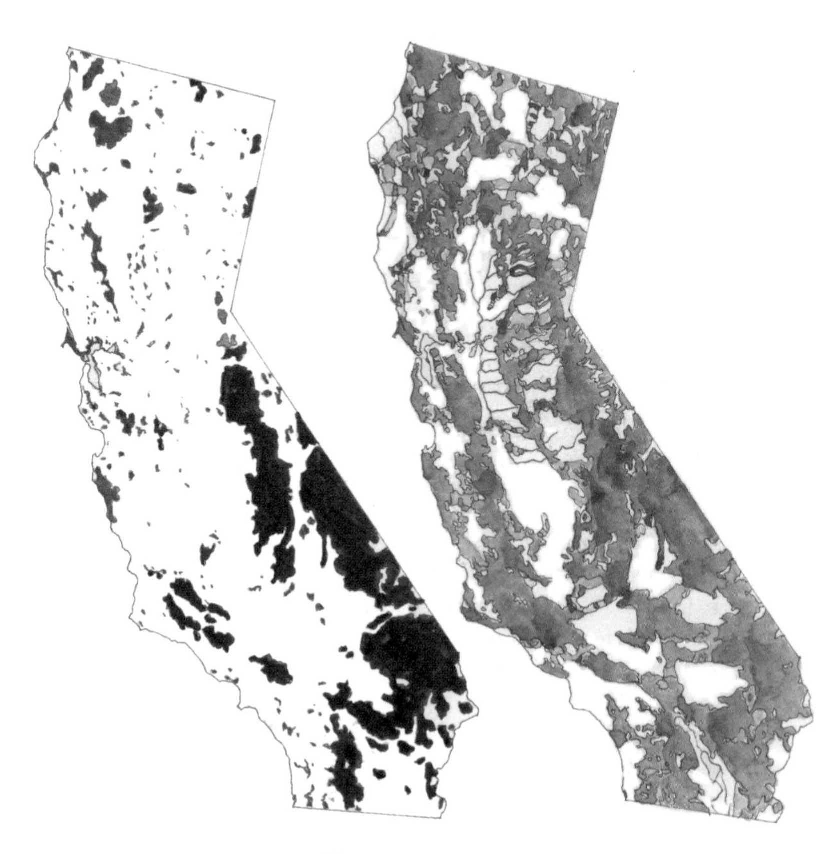

Map 03.04a

Map 03.04b

# 03.04 Repair, Restore, and Reciprocate

## *A vision of California in balance*

It is imaginable that the natural world of California can be stewarded to a state that is in better shape at the end of the twenty-first century than it was at the end of the twentieth. This means stewarding and restoring as much habitat as possible, especially corridors to connect chunks of protected landscape that are currently isolated. Now, the 2023 state law commonly referred to as 30×30 (SB337[1]) sets a baseline, prescribing a path forward to successfully maintain the biodiversity still extant across California.

The law calls for protecting 30 percent of California's land and coastal waters by the year 2030. The rhetoric in the 30×30 plan strongly indicates a desire to work with tribal governments across the state and to implement traditional ecological knowledge as a measure of the plan's success. But there is little mention of reinstating cultural burn patterns on publicly managed lands. The extent to which the plan makes good on its rhetoric may be a true measure of the 30×30 law's success in restoring fire resiliency and protecting tomorrow's landscapes. Regardless of criticisms, Sacramento's vision acknowledges a debt to be paid. An Indigenous critique of the industrial paradigm that has so fueled California's modern economy might call it, rather than a debt, a needed reciprocation. If so, reciprocation then is the currency of tomorrow's California, and the bill is due today. Decoupled from the extractive economies of the twentieth century, the economic juggernaut that is California may profit more from the protection and conservation of its resources than from their wanton destruction.

Map A. Policy—the current status of California's 30×30 law in terms of qualifying protected land designations: 24% terrestrial, 16% marine.[2] Qualifying lands are rated for their protection by being classified as either GAP 1 or GAP 2. GAP analysis, or the Gap Analysis Project, is a system of land designation that monitors the quality of biodiversity protection in terms of the land management plan in place to protect the land in question.[3]

The following are definitions of GAP 1 and GAP 2:[4]

* "GAP 1 land has a mandated management plan for biodiversity to prevent conversion of natural land cover and maintain a natural state. Natural disturbance events proceed or are mimicked in the management. Example: Wilderness Areas"

* "GAP 2 land has a mandated management plan for biodiversity to prevent conversion of natural land cover and maintain a natural state, but management practices can degrade natural states and natural disturbance events can be suppressed. Example: National Wildlife Refuges"

Map B. Essential Connectivity Map—based primarily on the concept of ecological integrity through designed corridors and improved connectivity to protect and bolster biodiversity.[5]

1. Irreplaceable and essential corridors
2. Linkages between habitat spaces, forming connective landscapes
3. Intact "natural" habitat areas
4. Limited connective opportunity—urban, agricultural, and developed areas

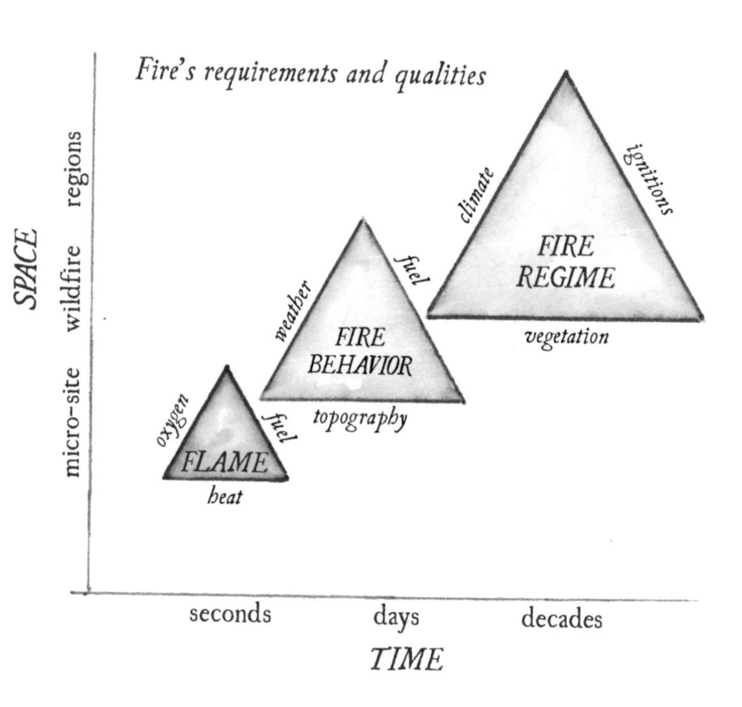
Fire's requirements and qualities
SPACE
micro-site
wildfire
regions
TIME
seconds
days
decades
FLAME
oxygen
fuel
heat
FIRE BEHAVIOR
weather
fuel
topography
FIRE REGIME
climate
ignitions
vegetation

# 03.05 Between Tradition and Innovation

## *A better story of fire*

In any ecosystem, much like in the human body, the ebb and flow of stress are a measure of health and resiliency. Because ecosystems don't die but succumb to type change, the analogy is limited, but the commonalities are interesting. Ecological stressors affect living systems on the land, and physiological processes in the body may include any combination of virulent factors that may or may not present themselves as events. Trespassing agents that work to disrupt the homeostatic health of the holobiont—the aggregate and integrated entity—also aid to maintain the holobiont's health by testing the parameters of its adaptive response and its capacity to recover. Health in the holobiont is more than a mechanical process. Health is a waveform function between stress and resiliency inside an ever-evolving process of maintaining equilibrium that is also a measure, in some very real sense, of sanity within the living system. The word *sane* is important and intentional. The recognizing of human cognitive processes—the functions of intelligence and consciousness—of something like a human mind inside the complexity of the ecological holobiont is more than an exercise in anthropomorphism. It is an exercise that establishes a kinship between different types of beings. If health is the measure of equilibrium, sanity is the measure of well-being. The kinship between the human and the more-than-human worlds forms a community of different types of selves, all individuals that collectively may negotiate ancient traditions and the many meanings of place and purpose.

*Fire is the mind of the land*
*thinking*
*botanically neurological pathways*
*designing*
*the health of the holobiont*

The story of fire has been at the heart of Californians' relationship to the land, long before the word *California* was uttered by any human mouth. The future story of fire in California has the potential to positively impact the lives of all Californians as much as any other opportunity ever could. As California moves from an age of progress into an age of resiliency, what fire cooperation by design may mean comes more and more into focus. Aspects of this brave new paradigm have the capacity to permeate all aspects of society and spur bold industries and sciences yet to be invented. The resurgence of traditional ecological knowledge, coupled with scientific innovation, will yield the strongest combination of tools to productively interact with the landscape that humanity could ask for. A cooperative relationship with fire has the potential to reform the California economy and to fuel a hundred new forms of industry that will all revolve around a centralized land ethic of reciprocation and intentionality. From architecture to health care, from agriculture to information technology, from real estate to outdoor recreation, from artificial intelligence to community gardening, all human relationships with the land in California are intimately connected to the land's relationship with fire. All private and public sectors of California's economy, including the drivers of energy distribution, logistics implementation, and communications infrastructure, will evolve in a rapidly changing climate, exacerbated by potentially bigger fire events and narrowing ecological bottlenecks. In a hotter and increasingly arid California, every industry will need to adapt to changing conditions; and with the advent of new technologies and the remembering of ancient stories, Californians have the tools at hand to get the job done. Just as pyrodiversity has informed biodiversity, fire informs human ecology and will increasingly inform economic development. Popular knowledge of why California burns is essential to the process. *You* are essential to the process.

As I realized in Whiskeytown, every ecosystem in the biosphere is now subject to humanity's legacy. In this age of climate and biodiversity crises, spawned in part from humanity's relationship with fire, feral wildland fire threateningly creeps toward the human sprawl everywhere all the time. Built patterns of urban and suburban growth form long trains of fuel for fire to express a kind of madness on the land, as if fire were on the hunt and will not be

denied. As homes burn and lives are lost because of decisions made and actions taken (or not taken), it is impossible not to wonder whether California has entered an age of inescapable and endless flame. In the quieter hours of winter and early spring, when the rains return, the green grass covers the hills, the data gets analyzed, consensus gets built, and the trauma from last year's fires begins to ebb, a vision of a possible alternate future comes into view. This vision of California's landscape and infrastructure made ever more resilient will be powered by a human community ready to tell stories again of resource reciprocation and abundance. This vision will be made real through the applied elegance of adaptive architectural, ecological, and political design that maximizes the integration of the natural and endemic processes of pyrogenic regeneration on the land and inside ourselves. The future of humanity's relationship with fire is dependent on humanity's relationship with itself. For every right that is celebrated, there is a corresponding responsibility to be met, and the measure of the collective will that rises to meet those responsibilities will gauge the future quality of humanity's continued residency in this most beautiful of all places, this perilous California.

when we look into the ephemeral flames
of a campfire, we see ourselves dancing
and reaching for the sky with what
potentially infinite fuel we are given,
and together, find ourselves worthy
of the gift.

# Acknowledgments

I've limited this list to colleagues, collaborators, and consultants with whom I've had direct conversations about the making of this book and its contents. In working on such a potentially contentious issue as fire in California, I've come to rely on the good-faith participation of the many brilliant minds who have directed, informed, and supported my work. This odd list doesn't include my family, my friends, or the larger community of booksellers, collectors, and supporters who make any part of this life I live beautiful, enjoyable, or even possible. The only way to truly express the gratitude, I feel, is to keep at it and to produce the best story I can. And yet I face the desperate realization that, in the end, I will never be able to repay the debt to so many who have supported me and this thing that I do. Thank you.

Adezio, Antonia / Artefact Design & Salvage
Alagona, Peter / *The Accidental Ecosystem*
Andersson, Mats / Indigofera
Anklam, Emmerich / Heyday
Asmuth, Paul / Meadowood Napa Valley
Bear, Lindsie / Humboldt Area Foundation
Bess, David / CDFW Law Enforcement Division
Brooks, Sarah / Trees Foundation
Burnett, Susan / Mojave Sands
Burns-Lambert, Laura / Baba Botanics
Caetano, Kalie / Heyday
Carle, David / David Carle Books
Carlon, John / River Partners
Carter, Richard / Carter and Co

Clapp, Heather / Point Reyes National Seashore Association
Clarke, Chris / *Ninety Miles to Needles*
Collins, Redgie / California Trout
Coryell, Julie / Literary publicist
Crowfoot, Wade / California Natural Resources Agency
Cueva, Nico / Haynes Vineyard
Dacey, Jessica / Mojave Desert Land Trust
Decker, Matt / Premium Arts
Dillon, Susan / California Association of Teachers of English
Dolman, Brock / Occidental Arts & Ecology Center
Drake, Jared / Wildbound PR
Drake, Julia / Wildbound PR
Dubois, Mark / Friends of the River
Einberger, Scott / National Park Service
Ferguson, Archie / Heyday
Fontaine, Dan / Wilderness Youth Project
Fullner, Michelle / *Golden State Naturalist*
Gamble, Launce / Gamble Ranch
George, Merv, Jr. / US Forest Service
Green, Mark / California Wilderness Coalition
Hanson, Chad / John Muir Project
Harris, Enich / Coast Film & Music Festival
Hart, Caryl / California Coastal Commission
Henson, Ryan / California Wilderness Coalition
Hinman, Daniela / Esselen Tribe
Hixon, Kindra / Mission Trails Regional Park Foundation
Hodge, Hilary / City of Grass Valley
Hoines, Josh / National Park Service
Hubbartt, Michael / *The Sutter Buttes*
Ingram, Ashley / Ashley Ingram Design
Ira, Gregory / UC Environmental Stewards
Jackson, Julia / *For the Wild*
Jones, Jeremy / Protect Our Winters

Jones, Michele / Copyeditor
Jung, Rebecca / Belvedere Tiburon Library
Karolyi, Alex / River Partners
Kauffmann, Michael / Backcountry Press
Kemp, Carroll / Alma Fria
Kenkel, Craig / National Park Service
Kishimoto, Yoriko / Anderson Valley Land Trust
Knight, Curtis / California Trout
Lambert, Roger / Black & Blue Ranch
Laws, John Muir / Wild Wonder Foundation
Margolin, Malcolm / California Institute for Community, Art, and Nature
Marshall, Jennifer / South Tahoe Public Utility District
Marson, Sabrina / Russian River Watershed Association
Martin, Manjula / *The Last Fire Season*
McCarron, Chris / Whiskeytown National Recreation Area
McElhatton, Mike / Anza-Borrego Desert Natural History Association
Merenlender, Adina / UC Environmental Stewards
Miller, Jeff / Center for Biological Diversity
Miranda, Deborah A. / *Bad Indians*
Miskwish, Michael Connolly / Sunbelt Publications
Mitchell, Shanley / Tuolumne River Trust
Montgomery, Peter / St. Dorothy's Rest
Morgan, Lance / Marine Conservation Institute
Morrill, Chris / California Wilderness Coalition
Murphy, Erma / O'Hanlon Center for the Arts
Nelson, Chad / Surfrider Foundation
Norris, Jennifer / California Natural Resources Agency
O'Brien, John P. / Mendocino Trail Stewards
O'Byrne, Eamon / Sonoma Land Trust
O'Donnell, Mackenzie / Mendocino Magic
O'Keeffe, Liv / California Native Plant Society
O'Neill, Timmy / *Soundscape*

Parlato, Liddy / Flowers Vineyards & Winery
Pipkin, Scot / Santa Barbara Botanic Garden
Pratt, Beth / National Wildlife Federation
Quintero, Armando / California State Parks
Raschke, Reenie / Garden Club of Montclair
Raz-Yaseef, Naama / Watershed Project
Reyes, Sammy / California State Parks
Rodgers, Andy / Russian River Watershed Association
Roebuck, Louesa / *Punk Ikebana*
Rothrock, Lily / River Partners
Saalisi, Dina / Dina Saalisi Healing Arts
Sarris, Greg / Federated Indians of Graton Rancheria
Satris, Marthine / Heyday
Schlickman, Emily / *Design by Fire*
Schneider, Charlie / Free the Eel
Sealy, Dan / Northcoast Environmental Center
Shaffer, Matthew / Sempervirens Fund
Sheppard, Laura / Mechanics' Institute
Smith, Terria / *News from Native California*
Solnit, Rebecca / *Orwell's Roses*
Stevenot, Austin / River Partners
Swimmer, Chad / Mendocino Trail Stewards
Tedesco, Marti / Peninsula Open Space Trust
Titus, Allison / Center for Natural Lands Management
Trinidad, Marcos / TreePeople
Vrooman, Lynette / Sierra College
Warner, Ben / Coast Film & Music Festival
Wasserman, Steve / Heyday
Wattawa, Gayle / Heyday
Wayburn, Laurie / Pacific Forest Trust
Wilder, Eric / Eric Wilder Graphics
Young, Ayana / Center for Humans and Nature

# Notes

Internet sources were accessed in 2023 and early 2024.

## PART ONE: FIRE HISTORY

### 01.01. Endemic/Indigenous Pyrosymbiosis


1. K. Aiyer, "The Great Oxidation Event: How Cyanobacteria Changed Life," American Society for Microbiology, February 18, 2022, https://asm.org/Articles/2022/February/The-Great-Oxidation-Event-How-Cyanobacteria-Change.

2. S. Bengtson, B. Rasmussen, M. Ivarsson, J. Muhling, C. Broman, F. Marone, M. Stampanoni, and A. Bekker, "Fungus-Like Mycelial Fossils in 2.4-Billion-Year-Old Vesicular Basalt," *Nature Ecology & Evolution* 1, no. 6 (2017): 0141, https://doi.org/10.1038/s41559-017-0141.

3. A. Scott, "The Pre-Quaternary History of Fire," *Palaeogeography, Palaeoclimatology, Palaeoecology* 164, nos. 1–4 (2000): 281–329, https://doi.org/10.1016/S0031-0182(00)00192-9.

4. B. Klein, O. Jagoutz, and J. Ramezani, "High-Precision Geochronology Requires That Ultrafast Mantle-Derived Magmatic Fluxes Built the Transcrustal Bear Valley Intrusive Suite, Sierra Nevada, California, USA," *Geology* 49, no. 1 (2020): 106–10, https://doi.org/10.1130/G47952.1.

5. Stanford University, "Ancient Raindrops Reveal the Origins of California's Sierra Nevada Range," *ScienceDaily*, July 9, 2006, www.sciencedaily.com/releases/2006/07/060709125422.htm.

6. A. Hipp, J. Cavender-Bares, and P. Manos, "Ascent of the Oaks," *Scientific American* 323, no. 2 (2020): 42, https://www.scientificamerican.com/article/how-oak-trees-evolved-to-rule-the-forests-of-the-northern-hemisphere.

7. S. Grimes, D. Mattey, J. Hooker, and M. Collinson, "Paleogene Paleoclimate Reconstruction Using Oxygen Isotopes from Land and Freshwater Organisms: The Use of Multiple Paleoproxies," *Geochimica et Cosmochimica Acta* 67, no. 21 (2003): 4033–47, https://doi.org/10.1016/S0016-7037(03)00173-X.

8. S. R. Holen, T. A. Demere, D. C. Fisher, R. Fullagar, J. B. Paces, G. T. Jefferson, J. M. Beeton, A. N. Rountrey, and K. A. Holen, "Disparate Perspectives on Evidence from the Cerutti Mastodon Site," *PaleoAmerica* 4, no. 1 (2018): 12–15, https://doi.org/10.1080/20555563.2017.1396836.

9. S. Pyne, *The Pyrocene* (Berkeley: University of California Press, 2021).

10. J. Vaillant, *Fire Weather* (New York: Knopf, 2023).

11. E. O. Wilson, *The Future of Life* (New York: Knopf, 2002).

12. J. Lovelock, *Gaia* (Oxford: Oxford University Press, 1979), 44.

## 01.02. Pyrodiversity and Biodiversity

1. R. E. Martin and D. B. Sapsis, "Fires as Agents of Biodiversity: Pyrodiversity Promotes Biodiversity," in *Proceedings of the Symposium on Biodiversity in Northwestern California, Oct. 1991*, ed. R. R. Harris and D. C. Erman, 150–57 (Berkeley: University of California, Division of Agricultural and Natural Resources, 1992).

2. G. M. Jones and M. W. Tingley, "Pyrodiversity and Biodiversity: A History, Synthesis, and Outlook," *Diversity and Distributions* 28, no. 3 (2022): 396–403, https://doi.org/10.1111/ddi.13280.

3. J. C. Hickman, *The Jepson Manual: Higher Plants of California* (Berkeley: University of California Press, 1993).

4. F. S. Chapin, B. H. Walker, R. J. Hobbs, D. U. Hooper, J. H. Lawton, O. E. Sala, and D. Tilman, "Biotic Control over the Functioning of Ecosystems," *Science* 270 (1997): 500–504.

5. B. J. Cardinale, J. E. Duffy, A. Gonzalez, D. U. Hooper, C. Perrings, P. Venail, A. Narwani, G. M. Mace, D. Tilman, D. A. Wardle, A. P. Kinzig, G. C. Daily, M. Loreau, J. B. Grace, A. Larigauderie, D. S. Srivastava, and S. Naeem, "Biodiversity: Loss and Its Impact on Humanity," *Nature* 486 (June 6, 2012): 59–67.

6. A. J. Huggett, "The Concept and Utility of 'Ecological Thresholds' in Biodiversity Conservation," *Biological Conservation* 124, no. 3 (2005): 301–10, https://doi.org/10.1016/j.biocon.2005.01.037.

7. T. H. Oliver, M. S. Heard, N.J.B. Isaac, D. B. Roy, D. Procter, F. Eigenbrod, R. Freckleton, A. Hector, C. David, L. Orme, O. L. Petchey, V. Proença, D. Raffaelli, K. B. Suttle, G. M. Mace, B. Martín-López, B. A. Woodcock, and J. M. Bullock, "Biodiversity and Resilience of Ecosystem Functions," *Trends in Ecology & Evolution* 30, no. 11 (2015): 673–84, https://doi.org/10.1016/j.tree.2015.08.009.

8. U. S. Rawat and N. K. Agarwal, "Biodiversity: Concept, Threats and Conservation," *Environment Conservation Journal* 16, no. 3 (2015): 19–28, https://doi.org/10.36953/ECJ.2015.16303.

9. C. Folke, S. Carpenter, T. Elmqvist, L. Gunderson, C. S. Holling, and B. Walker, "Resilience and Sustainable Development: Building Adaptive Capacity in a World of Transformations," *AMBIO: A Journal of the Human Environment* 31, no. 5 (2002): 437–40, https://doi.org/10.1579/0044-7447-31.5.437.

10. A. J. Mohammed, "Secretary-General's Video Message to Countdown to COP15: Leaders Event for a Nature-Positive World," September 20, 2022, https://www.un.org/sg/en/content/sg/statement/2022-09-20/secretary-generals-video-message-countdown-cop15-leaders-event-for-nature-positive-world.

11. N. G. Sugihara, J. W. Van Wagtendonk, and J. A. Fites-Kaufman, "Fire as an Ecological Process," in *Fire in California's Ecosystems*, 2nd ed., ed. J. W. van Wagtendonk, N. G. Sugihara, S. L. Stephens, A. E. Thode, K. E. Shaffer, and J. A. Fites-Kaufman, 57–70 (Berkeley: University of California Press, 2018).

12. USDA, Forest Service, Missoula Fire Sciences Laboratory, "Fire Regimes of California Chaparral Communities: Information from the Pacific Southwest Research Station and LANDFIRE," in Fire Effects Information System [online], 2018, USDA, Forest Service, Rocky Mountain Research Station, Missoula Fire Sciences Laboratory (Producer), www.fs.usda.gov/database/feis/fire_regimes/CA_chaparral/all.html.

13. J. E. Keeley, and F. W. Davis, "Chaparral," in *Terrestrial Vegetation of California*, 3rd ed., ed. M. G. Barbour, T. Keeler-Wolf, and

A. A. Schoenherr, 339–66 (Los Angeles: University of California Press, 2007).

## 01.03. A Tragedy of Misconceptions

1. C. Perry, *Pacific Arcadia: Images of California 1600–1915* (New York: Oxford University Press, 1999).

2. M. K. Anderson and M. J. Moratto, "Native American Land-Use Practices and Ecological Impacts," in *Sierra Nevada Ecosystem Project: Final Report to Congress*, vol. 2, *Assessments and Scientific Basis for Management Options*, 187–206 (Davis: University of California, Centers for Water and Wildland Resources, 1996).

3. V. H. Dale, S. Brown, R. A. Haeuber, N. T. Hobbs, N. Huntly, R. Naiman, W. Riebsame, M. Turner, and T. Valone, "Ecological Principles and Guidelines for Managing the Use of Land," *Ecological Applications* 10, no. 3 (2000): 639–70, https://doi.org/10.2307/2641032.

4. R. E. Martin and D. B. Sapsis, "Fires as Agents of Biodiversity: Pyrodiversity Promotes Biodiversity," in *Proceedings of the Symposium on Biodiversity in Northwestern California, Oct. 1991*, ed. R. R. Harris and D. C. Erman, 150–57 (Berkeley: University of California, Division of Agricultural and Natural Resources, 1992).

5. M. D. Abrams, "Fire in the Development of Oak Forests," *BioScience* 42, no. 5 (1992): 346–53.

6. T. R. Vale, "The Pre-European Landscape of the United States: Pristine or Humanized?" in *Fire, Native Peoples, and the Natural Landscape*, ed. T. R. Vale, 1–39 (Washington, DC: Island Press, 2002).

7. D. Graeber and D. Wengrow, *The Dawn of Everything: A New History of Humanity* (New York: Farrar, Straus and Giroux, 2021), 249–75.

8. R. L. Bettinger, *Orderly Anarchy: Sociopolitical Evolution in Aboriginal California* (Berkeley: University of California Press, 2015).

9. James Rust, Southern Miwok elder, quoted in M. K. Anderson, *Tending the Wild* (Berkeley: University of California Press, 2005), 3.

10. "Wallace Stegner," n.d., Wilderness Society, https://www.wilderness.org/articles/article/wallace-stegner.

11. G. Sarris, *How a Mountain Is Made* (Berkeley: Heyday, 2017), 272.

12. G. M. Sanchez, M. Grone, and A. Apodaca, "Indigenous Stewardship of Coastal Resources in Native California," *Frontiers in Earth Science* 11 (2023), https://doi.org/10.3389/feart.2023.1064197.

13. M. Turco, J. T. Abatzoglou, S. Herrera, Y. Zhuang, S. Jerez, D. D. Lucas, A. AghaKouchak, and I. Cvijanovic, "Anthropogenic Climate Change Impacts Exacerbate Summer Forest Fires in California," *PNAS* 120, no. 25 (2023): e2213815120, https://doi.org/10.1073/pnas.2213815120.

14. M. Willeit, A. Ganopolski, R. Calov, A. Robinson, and M. Maslin, "The Role of CO2 Decline for the Onset of Northern Hemisphere Glaciation," *Quaternary Science Reviews* 119 (2015): 22–34, https://doi.org/10.1016/j.quascirev.2015.04.015.

15. E. Anters, "Precipitation and Water Supply in the Sierra Nevada, California," *Bulletin of the American Meteorological Society* 20 (1938): 89–91.

16. L. V. Benson, M. Kashgarian, R. Rye, S. Lund, F. Paillet, J. Smoot, C. Kester, S. Mensing, D. Meko, and S. Lindstrom, "Holocene Multidecadal and Multicentennial Droughts Affecting Northern California and Nevada," *Quaternary Science Reviews* 21, nos. 4–6 (2002): 659–82, https://doi.org/10.1016/S0277-3791(01)00048-8.

17. K. Lightfoot, "Cultural Construction of Coastal Landscapes: A Middle Holocene Perspective from San Francisco Bay," in *Archaeology of the California Coast during the Middle Holocene*, ed. J. Erlanson and M. Glassow, 129–41 (Los Angeles: Cotsen Institute of Archaeology, 1997).

18. R. S. Bradley, "1000 Years of Climate Change," *Science* 288, no. 5470 (2000): 1353–44, https://doi.org/10.1126/science.288.5470.1353.

19. A. Schimmelmann, C. B. Lange, and B. L. Meggers, "Paleoclimatic and Archaeological Evidence for a 200-Year Recurrence of Floods and Droughts Linking California, Mesoamerica and South America over the Past 2000 Years," *Holocene* 13, no. 5 (2003): 763–78, https://doi.org/10.1191/0959683603hl661rp.

20. S. A. Mensing, R. Byrne, and J. Michaelsen, "A 560-Year Record of Santa Ana Fires Reconstructed from Charcoal Deposited in the Santa Barbara Basin, California," *Quaternary Research* 51, no. 3 (1999): 295–305, https://doi.org/10.1006/qres.1999.2035.

21. P. H. Gleick and M. Heberger, "Devastating Drought Seems Inevitable in the American West," *Scientific American*, January 8, 2012, https://www.scientificamerican.com/article/the-coming-mega-drought.

22. Lightfoot, "Cultural Construction of Coastal Landscapes."

## 01.04. Cultural Fire on the Land

1. L. M. Raab and T. L. Jones, "The Rediscovery of California Prehistory," in *Prehistoric California: Archaeology and the Myth of Paradise*, ed. L. M. Raab and T. L. Jones, 1–10 (Salt Lake City: University of Utah Press, 2002).

2. G. J. Nowacki, D. W. MacClerry, and F. K. Lake, "Native Americans, Ecosystem Development, and Historical Range of Variation," in *Historical Environmental Variation in Conservation and Natural Resource Management*, ed. J. A. Weins, 76–91 (Hoboken, NJ: Wiley-Blackwell, 2012).

3. S. L. Stephens, R. E. Martin, and N. E. Clinton, "Prehistoric Fire Area and Emissions from California's Forests, Woodlands, Shrublands, and Grasslands," *Forest Ecology and Management* 251, no. 3 (2007): 205–16, https://doi.org/10.1016/j.foreco.2007.06.005.

4. S. A. Mensing, R. Byrne, and J. Michaelsen, "A 560-year Record of Santa Ana Fires Reconstructed from Charcoal Deposited in the Santa Barbara Basin, California," *Quaternary Research* 51, no. 3 (1999): 295–305, https://doi.org/10.1006/qres.1999.2035.

5. J. E. Keeley, "Native American Impacts on Fire Regimes of the California Coastal Ranges," *Journal of Biogeography* 29, no. 3 (2002): 303–20, https://www.jstor.org/stable/827540.

6. C. L. Crumley, "Historical Ecology: A Multidimensional Ecological Orientation," in *Historical Ecology: Cultural Knowledge and Changing Landscapes*, ed. C. L. Crumley (Santa Fe, NM: School of American Research Press, 1994).

7. G. P. Nabhan and A. Rea, "Plant Domestication and Folk-Biological Change: The Upper Piman/Devil's Claw Example," *American Anthropologist* 89, no. 1 (1987): 57–73.

8. R. Ornduff, *An Introduction to California Plant Life* (Berkeley: University of California Press, 1974).

9. M. K. Anderson, "Maintaining and Extending the Coastal Prairies," in *Tending the Wild*, 165–185 (Berkeley: University of California Press, 2005).

10. J. K. Agee, *Fire Ecology of Pacific Northwest Forests* (Washington, DC: Island Press, 1993).

11. J. H. Connell, "Diversity in Tropical Rainforests and Coral Reefs," *Science* 199, no. 4335 (1978): 1302–10, https://doi.org/10.1126/science.199.4335.1302.

12. M. K. Anderson, "Indian Fire-Based Management in the Sequoia-Mixed Conifer Forests of the Central and Southern Sierra Nevada," final report to the Yosemite Research Center, Yosemite National Park, Cooperative Agreement Order Number 8027-002 (San Francisco: US Department of the Interior, National Park Service, Western Region, 1993).

## 01.05. Colonial Conflagration

1. E. Bruenig, "A Saint's Sins: In California, Protesters Have Toppled Statues of Junipero Serra, Whose Missions Brutalized Native Americans. How Should We Think of Him Now?" *New York Times*, August 16, 2020, https://www.nytimes.com/2020/08/16/opinion/junipero-serra-catholic-saint.html.

2. D. Krell, ed., *The California Missions: A Pictorial History* (Menlo Park, CA: Sunset Publishing, 1979).

3. L. T. Burcham, "Cattle and Range Forage in California: 1770–1880," *Agricultural History* 35, no. 3 (1961): 140–49, http://www.jstor.org/stable/3740625.

4. K. G. Lightfoot, *Indians, Missionaries, and Merchants: The Legacy of Colonial Encounters on the California Frontiers* (Berkeley: University of California Press, 2005).

5. B. Madley, *An American Genocide: The United States and the California Indian Catastrophe* (New Haven, CT: Yale University Press, 2017).

6. E. M. Johnson, "How John Muir's Brand of Conservation Led to the Decline of Yosemite," *Scientific American*, August 13, 2014, https://blogs.scientificamerican.com/primate-diaries/how-john-muir-s-brand-of-conservation-led-to-the-decline-of-yosemite.

7. CAL FIRE, "Our Organization," n.d., https://www.fire.ca.gov/about/our-organization.

8. G. Pinchot, "The Relation of Forests and Forest Fires," *National Geographic* 10 (1899), https://foresthistory.org/wp-content/uploads/2016/12/pinchot_Relation-of-forests-and-forest-fires.pdf.

9. T. Egan, *The Big Burn: Teddy Roosevelt and the Fire That Saved America* (Boston: Houghton Mifflin Harcourt, 2009), 239–48.

10. E. Goodman, "1970 Law Gives Newsom Sweeping Powers in a State of Emergency; Local Official Helped Write the Law," *Palo Alto Daily Post*, May 21, 2020, https://padailypost.com/2020/05/21/1970-law-gives-newsom-sweeping-powers-in-a-state-of-emergency-local-official-helped-write-the-law.

11. Tall Timbers, "Mission and Philosophy," n.d., https://talltimbers.org/mission-and-philosophy.

12. S. Pyne, "Two Contagions, One Opportunity to Reboot Our Approach," *History News Network*, July 26, 2020, https://historynews-network.org/article/176572.

13. Los Angeles Firemen's Relief Association, "The Bel Air Fire, November 6, 1961—Revisited," October 31, 2015, https://www.lafra.org/lafd-history-the-bel-air-fire-november-6-1961-revisited.

14. Wallace Stegner, "Wilderness Letter," 1964, https://web.stanford.edu/&cbross/Ecospeak/wildernessletter.html stanford.edu.

15. J. Wilkens, "California Was on Fire 50 Years Ago, Too," *San Diego Union Tribune*, August 30, 2020, https://www.sandiegouniontribune.com/news/public-safety/story/2020-08-30/california-fires-1970-legacy.

16. D. Sapsis and T. Moody, "Indicators of Climate Change in California," CAL FIRE report 04wildfires.pdf, California Office of Environmental Health Hazard Assessment, 2022, https://oehha.ca.gov/media/downloads/climate-change/document/04wildfires.pdf.

## 01.06. The Coming of Modern Megafire


1. National Geographic, s. v. "Megafire," accessed February 21, 2024, https://education.nationalgeographic.org/resource/megafire.
2. "Largest California Wildfires," Reuters Graphics, accessed March 7, 2024, https://www.reuters.com/graphics/CALIFORNIA-WILDFIRES/gdpzyjxmovw/.
3. C. Miller and D. L. Urban, "Forest Pattern, Fire, and Climatic Change in the Sierra Nevada," *Ecosystems* 2 (1999): 76–87, https://doi.org/10.1007/s100219900060.
4. A. L. Westerling, H. G. Hidalgo, D. R. Cayan, and T. W. Swetnam, "Warming and Earlier Spring Increase Western U.S. Forest Wildfire Activity," *Science* 313, no. 5789 (2006): 940–43, https://doi.org/10.1126/science.1128834.
5. C. J. Fettig, S. M. Hood, J. B. Runyon, and C. M. Stalling, "Bark Beetle and Fire Interaction in Western Coniferous Forests," *Fire Management Today* 79, no. 1 (2021): 14–23, https://www.fs.usda.gov/psw/publications/fettig/psw_2021_fettig006.pdf.
6. J. Keeley, J. Franklin, and C. D'Antonio, "Fire and Invasive Plants on California Landscapes," in *The Landscape Ecology of Fire*, ed. D. McKenzie, C. Miller, and D. A. Falk, 193–221 (Dordrecht, Netherlands: Springer, 2011).
7. H. D. Safford, "Man and Fire in Southern California: Doing the Math," *Fremontia* 35, no. 4 (2007): 25–29.
8. B. M. Collins, J. D. Miller, E. E. Knapp, and D. B. Sapsis, "A Quantitative Comparison of Forest Fires in Central and Northern California under Early (1911–1924) and Contemporary (2002–2015) Fire Suppression," *International Journal of Wildland Fire* 28 (2019): 138–48, https://doi.org/10.1071/WF18137.
9. J. E. Keeley and A. D. Syphard, "Different Fire-Climate Relationships on Forested and Non-Forested Landscapes in the Sierra Nevada Ecogregion," *International Journal of Wildland Fire* 24 (2015): 27–36, https://doi.org/10.1071/WF14102.

10. E. Loverridge, "The Fire Suppression Policy of the U.S. Forest Service," *Journal of Forestry* 42, no. 8 (1944): 549–54, https://doi.org/10.1093/jof/42.8.549.

11. M. K. Hughes and P. M. Brown, "Drought Frequency in Central California since 101 BC Recorded in Giant Sequoia Tree Rings," *Climate Dynamics* 6 (1992): 161–67, https://doi.org/10.1007/BF00193528.

12. J. Farmer, "The Golden State Treescape Wasn't Made to Last," *Los Angeles Times*, November 1, 2020, https://www.latimes.com/opinion/story/2020-11-01/california-wildfire-forest-sequoias-eucalyptus-climate-change.

13. H. W. Debruin, "From Fire Control to Fire Management: A Major Policy Change in the Forest Service," *Proceedings of the Tall Timbers Fire Ecology Conference* 14 (1974): 11–17, https://talltimbers.org/wp-content/uploads/2014/03/DeBruin1974_op.pdf.

## 01.07. The Century of Reset and Reckoning

1. "Largest California Wildfires," Reuters Graphics, accessed March 7, 2024, https://www.reuters.com/graphics/CALIFORNIA-WILDFIRES/gdpzyjxmovw/.

2. National Wildfire Coordinating Group, "Glossary of Wildland Fire Terminology," PMS 205, July 2012, https://www.nwcg.gov/sites/default/files/data-standards/glossary/pms205.pdf.

3. USDA, Forest Service, "Wildland Fire," n.d., https://www.fs.usda.gov/managing-land/fire.

4. G. J. Williamson, T. M. Ellis, and D. M. Bowman, "Double-Differenced dNBR: Combining MODIS and Landsat Imagery to Map Fine-Grained Fire MOSAICS in Lowland Eucalyptus Savanna in Kakadu National Park, Northern Australia," *Fire* 5, no. 5 (2022): 160, https://doi.org/10.3390/fire5050160.

# PART TWO: FIRE ECOLOGY

## 02.01. Cascading Patterns of Emergence

1. Union of Concerned Scientists, "The Science of Forest Fires: Culture, Climate, and Combustion," July 23, 2018, https://www.ucsusa.org/resources/science-forest-fires-culture-climate-and-combustion.

2. Examples: the Department of Fire Ecology Research at the University of Nevada, Reno (https://www.unr.edu/nres/research/fire); the Department of Fire Ecology at Oregon State University (https://www.forestry.oregonstate.edu/undergraduate-programs/natural-resources/wildland-fire-ecology); Northern Arizona University (https://news.nau.edu/forestry-continuing-education/); and Arizona University's Fire Ecology Department (https://nature.arizona.edu/research/fire-ecology).

3. E. P. Odum, "The Strategy of Ecosystem Development," *Science* 164, no. 3877 (1969): 262–70, https://doi.org/10.1126/science.164.3877.262.

4. A. G. Tansley, "The Use and Abuse of Vegetational Concepts and Terms," *Ecology* 16, no. 3 (1935): 196–218, https://doi.org/10.2307/1930070.

5. N. G. Sugihara, J. W. Van Wagtendonk, and J. A. Fites-Kaufman, "Fire as an Ecological Process," in *Fire in California's Ecosystems*, 2nd ed., ed. J. W. van Wagtendonk, N. G. Sugihara, S. L. Stephens, A. E. Thode, K. E. Shaffer, and J. A. Fites-Kaufman, 57–70 (Berkeley: University of California Press, 2018).

6. A. M. Schultz, "The Ecosystem as a Conceptual Tool in the Management of Natural Resources," in *Natural Resources Quantity and Quality*, ed. S. V. Cirancy-Wantrup and J. J. Parsons, 139–61 (Berkeley: University of California Press, 1968).

7. Z. L. Steel, M. J. Koontz, and H. D. Safford, "The Changing Landscape of Wildfire: Burn Pattern Trends and Implications for California's Yellow Pine and Mixed Conifer Forests," *Landscape Ecology* 33 (2018): 1159–76, https://doi.org/10.1007/s10980-018-0665-5.

8. C. Tymstra, "A Burning Question," *Metascience* 32 (2023): 413–16, doi.org/10.1007/s11016-023-00866-5.

9. R. E. Brazier, A. Puttock, H. A. Graham, R. E. Auster, K. H. Davies, and C. M. Brown, "Beaver: Nature's Ecosystem Engineers," *WIREs Water* 8, no. 1 (2021): e1494, https://doi.org/10.1002/wat2.1494.

10. J. Cleveland, "Firefighting Beavers," USDA, Forest Service, October 2, 2023, https://www.fs.usda.gov/features/firefighting-beavers.

11. J. K. Wood, N. Nadav, C. A. Howell, and G. R. Geupel, "Overview of Cosumnes Riparian Bird Study and Recommendations for Monitoring and Management," report to the California Bay-Delta Authority Ecosystem Restoration Program, Contract ERP-01-NO1 PRBO, 2006, https://watershed.ucdavis.edu/sites/g/files/dgvnsk8531/files/products /2022-04/PRBO_Overview_Woodetal2006.pdf.

12. M. E. Seamans and R. J. Gutierrez, "Habitat Selection in a Changing Environment: The Relationship between Habitat Alteration and Spotted Owl Territory Occupancy and Breeding Dispersal," *Condor* 109, no. 3 (2007): 566–76.

13. D. W. Johnson, R. B. Susfalk, T. G. Caldwell, J. D. Murphy, W. W. Miller, and R. F. Walker, "Fire Effects on Carbon and Nitrogen Budgets in Forests," *Water, Air, and Soil Pollution* 4 (2004): 263–75, https://doi.org/10.1023/B:WAFO.0000028359.17442.d1.

14. G. Giovaninni and S. Lucchesi, "Modifications Induced in Soil Physico-Chemical Parameters by Experimental Fires at Different Intensities," *Soil Science* 162, no. 7 (1997): 479–86, https://doi.org/10.1097 /00010694-199707000-00003.

15. P. M. Wohlgemuth, K. Hubbert, T. Procter, and S. Ahuja, "Fire and Physical Environment Interactions: Soil, Water, and Air," in *Fire in California's Ecosystems*, 87–102 (Berkeley: University of California Press, 2018).

16. A. R. Tiedemann, C. E. Conrad, J. H. Dieterich, J. W. Hornbeck, W. F. Megahan, L. A. Viereck, and D. D. Wade, *Effects of Fire on Water: A State-of-Knowledge Review*, General Technical Report WO-GTR-10 (Washington, DC: USDA, Forest Service, 1979).

17. S. L. Gutsell and E. A. Johnson, "How Fire Scars Are Formed: Coupling a Disturbance Process to Its Ecological Effect," *Canadian Journal of Forest Research* 26, no. 2 (1996): 166–174, https://doi.org/10.1139 /x26-020.

18. M. R. Gallagher, J. K. Kreye, E. T. Machtinger, A. Everland, N. Schmidt, and N. S. Skowronski, "Can Restoration of Fire-Dependent Ecosystems Reduce Ticks and Tick-Borne Disease Prevalence in the Eastern United States?" *Ecological Applications* 32, no. 7 (2022): e2637, https://doi.org/10.1002/eap.2637.

19. K. Fisher, K. M. Watrous, N. M. Williams, L. L. Richardson, and S. H. Woodard, "A Contemporary Survey of Bumble Bee Diversity across the State of California," *Ecology and Evolution* 12 (2022): e8505, https://doi.org/10.1002/ece3.8505.

20. J. W. Rivers and M. G. Betts, "Postharvest Bee Diversity Is High but Declines Rapidly with Stand Age in Regenerating Douglas-Fir Forest," *Forest Science* 67, no. 3 (2021): 275–85, https://doi.org/10.1093/forsci/fxab002.

21. A. Fidelis, L. Zirondi, and L. Heloiza, "And after Fire, the Cerrado Flowers: A Review of Post-Fire Flowering in a Tropical Savanna," *Flora* 280 (2021): 151849, https://doi.org/10.1016/j.flora.2021.151849.

22. V. Wojcik, "Pollinators: Their Evolution, Ecology, Management, and Conservation," in *Arthropods: Are They Beneficial for Mankind?* ed. R. E. Ranz (London: Intechopen, 2021), https://doi.org/10.5772/intechopen.97153.

23. J. Hanula, S. Horn, and J. O'Brien, "Have Changing Forests Conditions Contributed to Pollinator Decline in the Southeastern United States?" *Forest Ecology and Management* 348 (2015): 142–52, https://doi.org/10.1016/j.foreco.2015.03.044.

24. J. Hillerislambers, S. G. Yelenik, B. P. Colman, and J. M. Levine, "California Annual Grass Invaders: The Drivers or Passengers of Change?" *Journal of Ecology* 98, no. 5 (2010): 1147–56, https://doi.org/10.1111/j.1365-2745.2010.01706.x.

25. B. Sandel and E. Dangremond, "Climate Change and the Invasion of California by Grasses," *Global Change Biology* 18, no. 1 (2012): 277–89, https://doi.org/10.1111/j.1365-2486.2011.02480.x.

26. S. M. Vallina and C. Le Quéré, "Stability of Complex Food Webs: Resilience, Resistance and the Average Interaction Strength," *Journal of Theoretical Biology* 272, no. 1 (2011): 160–73, https://doi.org/10.1016/j.jtbi.2010.11.043.

## 02.02. Succession and Conversion

1. J. E. Lovelock, *Gaia: A New Look at Life on Earth* (Oxford: Oxford University Press, 1979).
2. S. M. Sundstrom and C. R. Allen, "The Adaptive Cycle: More Than a Metaphor," *Ecological Complexity* 39 (2019), 100767, https://doi.org/10.1016/j.ecocom.2019.100767.
3. A. M. Schults, "The Ecosystem as a Conceptual Tool in the Management of Natural Resources," in *Natural Resources: Quality and Quantity*, ed. S. V. Cirancy-Wantrup and J. J. Parsons, 139–61 (Berkeley: University of California Press, 1968).
4. J. K. Agee, *Fire Ecology of Pacific Northwest Forests* (Washington, DC: Island Press, 1993).
5. J. O. Sawyer, T. Keeler-Wolf, and J. M. Evens, *A Manual of California Vegetation*, 2nd ed. (Sacramento: California Native Plant Society, 2009).
6. Fire and Resource Assessment Program of the California Department of Forestry and Fire Protection, *California's Forests and Rangelands: 2017 Assessment*, CAL FIRE, 2018, https://cdnverify.frap.fire.ca.gov/media/4babn5pw/assessment2017.pdf.

## 02.03. Patterns of Vulnerability and Resilience

1. K. L. Calhoun, M. Chapman, C. Tubbesing, A. McInturff, K. Gaynor, A. Van Scoyoc, C. E. Wilkinson, P. Parker-Shames, D. Kurz, and J. Brashares, "Spatial Overlap of Wildfire and Biodiversity in California Highlights Gap in Non-Conifer Fire Research and Management," *Diversity and Distributions* 28, no. 3 (2022): 529–41, https://doi.org/10.1111/ddi.13394.
2. J. E. Keeley and H. D. Safford, "Fire as an Ecosystem Process," in *Ecosystems of California*, ed. H. Mooney and E. Zavaleta, 27-45 (Oakland: University of California Press, 2016).
3. J. W. van Wagtendonk and D. R. Cayan, "Temporal and Spatial Distribution of Lightning Strikes in California in Relationship to Large-Scale Weather Patterns," *Fire Ecology* 4, no. 1 (2008): 34–56, https://doi.org/10.4996/fireecology.0401034.

4. S. R. Miles and C. B. Goudey, comps., with major contributions by E. B. Alexander and J. O. Sawyer, *Ecological Subregions of California: Section and Subsection Descriptions*, Technical Publication R5-EM-TP-005 (San Francisco: USDA, Forest Service, Pacific Southwest Region, 1997).

5. H. T. Lewis, "Patterns of Indian Burning in California: Ecology and Ethnohistory," in *Before the Wilderness: Environmental Management by Native Californians*, ed. T. C. Blackburn, 56–116 (Menlo Park, CA: Ballena Press, 1993).

6. G. J. West, "The Late Pleistocene-Holocene Pollen Record and Prehistory of California's North Coast Ranges," in *There Grows a Green Tree: Papers in Honor of David A. Fredrickson*, ed. G. White, P. Mikkelson, M. Hildebrandt, and M. Basgall, 65–80 (Davis: University of California, 1993).

7. D. A. Hatch, J. W. Bartolome, J. S. Fehmi, and D. S. Hillyard, "Effects of Burning or Grazing on a Coastal California Grassland," *Restoration Ecology* 7, no. 4 (1999): 376–81, https://doi.org/10.1046/j.1526-100X.1999.72032.x.

8. C. A. Zammit and P. H. Zedler, "The Influence of Dominant Shrubs, Fire, and Time since Fire on Soil Seed Banks in Mixed Chaparral," *Vegetation* 75 (1988): 175–87, https://doi.org/10.1007/BF00045632.

9. A. B. Forrestel, M. A. Moritz, and S. L. Stephens, "Landscape-Scale Vegetation Change following Fire in Point Reyes, CA," *Fire Ecology* 7, no. 2 (2011): 114–28, https://doi.org/10.4996/fireecology.0702114.

10. B. J. Harvey and B. A. Holzman, "Divergent Successional Pathways of Stand Development Following Fire in a California Closed-Cone Pine Forest," *Journal of Vegetation Science* 25, no. 1 (2013): 88–99, https://doi.org/10.1111/jvs.12073.

11. L. Narayan, "Clonal Diversity, Patterns, and Structure in Old Coast Redwood Forests," UC Berkeley, 2015. https://escholarship.org/uc/item/74j6599q.

12. G. King, *The Ghost Forest: Racists, Radicals, and Real Estate in the California Redwoods* (New York: Public Affairs, 2023).

13. G. Snyder, *Mountains and Rivers without End* (Berkeley, CA: Counterpoint, 1996).

14. C. E. Briles, C. Whitlock, C. N. Skinner, and J. Mohr, "Holocene Forest Development and Maintenance on Different Substrates in the Klamath Mountains, Northern California, USA," *Ecology* 92, no. 3 (2011): 590–601, https://doi.org/10.1890/09-1772.1.

15. T. Atzet and R. Martin, "Natural Disturbance Regimes in the Klamath Province," in *Proceedings of the Symposium on Biodiversity in Northwestern California, Oct. 1991*, ed. R. R. Harris and D. C. Erman, 40–48 (Berkeley: University of California, Division of Agricultural and Natural Resources, 1992).

16. J. K. Agee, *Fire Ecology of Pacific Northwest Forests* (Washington, DC: Island Press, 1991).

17. S. L. Stephens, and M. A. Finney, "Prescribed Fire Mortality of Sierra Nevada Mixed Conifer Tree Species: Effects on Crown Damage and Forest Floor Combustion," *Forest Ecology and Management* 162, nos. 2–3 (2002): 261–71, https://doi.org/10.1016/S0378-1127(01)00521-7.

18. J. Woodward and F. C. Coe, "Wildlife in Managed Forests: Fisher and Humboldt Martin," Oregon Forest Resources Institute, 2018, https://oregonforests.org/sites/default/files/2018-12/ManagedForests_Carnivores_2018-WEB.pdf.

19. US Geological Survey, "Mount Shasta," California Volcano Observatory, accessed March 8, 2024, https://www.usgs.gov/volcanoes/mount-shasta.

20. W. W. Oliver, "Can We Create and Sustain Late Successional Attributes in Interior Ponderosa Pine Stands? Large-Scale Ecological Research Studies in Northeastern California," in *Ponderosa Pine Ecosystems Restoration and Conservation: Steps toward Stewardship*, USDA Forest Service Proceedings RMRS-P22, compiled by R. K. Vance, C. B. Edminster, W. W. Covington, and J. A. Blake, 99–103 (Ogden, UT: Rocky Mountain Research Station, 2001), https://www.fs.usda.gov/rm/pubs/rmrs_p022/rmrs_p022_099_103.pdf.

21. J. K. Agee, "Fire Regimes and Approaches for Determining Fire History," in *The Use of Fire in Forest Restoration*, General Technical Report INT-GTR-341, ed. C. G. Hardy and S. F. Arno, 12–13 (Ogden, UT: USDA, Forest Service, Intermountain Research Station, 1996).

22. Lassen Volcanic National Park and Yosemite National Park, "Tracking One of California's Rarest Mammals," National Park Service, 2018, https://www.nps.gov/articles/tracking-snrf.htm.

23. A. H. Winward, "Fire in the Sagebrush-Grass Ecosystems," in *Rangeland Fire Effects*, ed. K. Sanders and J. Durham, 2–6 (Boise, ID: USDI Bureau of Land Management, 1985).

24. C. C. Frost, "Presettlement Fire Frequency Regimes of the United States: A First Approximation," *Proceedings Tall Timbers Fire Ecology Conference* 20 (1998): 70–81.

25. Fire and Resource Assessment Program of the California Department of Forestry and Fire Protection, *California's Forests and Rangelands: 2017 Assessment*, CAL FIRE, 2018, https://cdnverify.frap.fire.ca.gov/media/4babn5pw/assessment2017.pdf.

26. A. Larson, "2023 Is 2nd-Snowiest Winter Recorded in Sierra Nevada Mountains," KRON TV, March 21, 2023, https://www.kron4.com/news/california/2023-is-2nd-snowiest-winter-recorded-in-sierra-nevada-mountains.

27. C. C. Lee, "Weather Whiplash: Trends in Rapid Temperature Changes in a Warming Climate," *International Journal of Climatology* 42, no. 8 (2022), 4214–22, https://doi.org/10.1002/joc.7458.

28. Fire and Resource Assessment Program of the California Department of Forestry and Fire Protection, *California's Forests and Rangelands*.

29. Natural Resources Conservation Service, *Inventory Summary Report* (Ames: Iowa State University, USDA Center for Survey Statistics and Methodology, 2012).

30. B. H. Allen-Diaz and J. W. Bartolome, "Survival of *Quercus douglasii* Seedlings under the Influence of Fire and Grazing," *Madrono* 39, no. 1 (1992): 47–53, https://www.jstor.org/stable/41424885.

31. S. Barry, S. Larson, and M. George, "California Native Grasslands: A Historical Perspective, a Guide for Developing Realistic Restoration Objectives," *Grasslands* 30, no. 1 (Winter 2020), https://cnga.org/resources/Documents/Grasslands%20Journal/Grassland%20Issues/2020%20Grasslands%20Journal/CNGA%20Grasslands%202020%20Vol%2030%20No%201.pdf.

32. H. H. Biswell, "Ecology of California Grasslands," *Journal of Range Management* 9 (1956): 19–24, https://ucanr.edu/repository/fileaccess .cfm?article=157150&p=ZHIFXI .

33. L. M. Ellis, "Floods and Fire along the Rio Grande: The Role of Disturbance in the Riparian Forest" (PhD diss., University of New Mexico, 1999).

34. Office of the Governor, "California Releases Beavers into the Wild for First Time in Nearly 75 Years," December 13, 2023, https://www.gov. ca.gov/2023/12/13/california-releases-beavers-into-the-wild-for-first -time-in-nearly-75-years.

35. D. T. Fischer, C. J. Still, and A. P. Williams, "Significance of Summer Fog and Overcast for Drought Stress and Ecological Functioning of Coastal California Endemic Plant Species," *Journal of Biogeography* 36, no. 4 (2009):783–99, https://doi.org/10.1111/j.1365 -2699.2008.02025.x.

36. M. A. Moritz, "Analyzing Extreme Disturbance Events: Fire in Los Padres National Forest," *Ecological Applications* 7, no. 4 (1997): 1252–62, https://doi.org/10.1890/1051-0761(1997)007[1252:AEDEFI _2.0.CO;2.

37. B. P. Bryant and A. L. Westerling, "Scenarios for Future Wildfire Risk in California: Links between Changing Demography, Land Use, Climate, and Wildfire," *Environmetrics* 25, no. 6 (2014): 454–71, https://doi.org/10.1002/env.2280.

38. California Department of Fish and Wildlife, "Tule Elk," n.d., https:// wildlife.ca.gov/conservation/mammals/elk/tule-elk#341091208 -distribution--range.

39. F. W. Davis, and J. Michaelsen, "Sensitivity of Fire Regime in Chaparral Ecosystems to Climate Change," in *Global Change and Mediterranean-Type Ecosystems*, ed. J. M. Moreno and W. C. Oechel, 435–56 (New York: Springer-Verlag, 1995).

40. D. R. Weise, J. C. Regelbrugge, T. E. Paysen, and S. G. Conard, "Fire Occurrence on Southern California National Forests—Has It Changed Recently?" in *Proceedings from the Symposium: Fire in California Ecosystems: Integrating Ecology, Prevention and Management*,

Miscellaneous Publication No. 1, ed. N. G. Sugihara, M. E. Morales, and T. J. Morales, 389–91 (Sacramento: Association for Fire Ecology, 2002).

41. J. E. Keeley and P. H. Zedler, "Large, High-Intensity Fire Events in Southern California Shrublands: Debunking the Fine-Grained Age-Patch Model," *Ecological Applications* 19, no. 1 (2009): 69–94, https://doi.org/10.1890/08-0281.1.

42. L. L. Esser, "*Pinus torreyana*," in Fire Effects Information System [online], 1993, US Department of Agriculture, Forest Service, Rocky Mountain Research Station, Fire Sciences Laboratory (Producer), https://www.fs.usda.gov/database/feis/plants/tree/pintor/all.html.

43. S. G. Whisenant, "Changing Fire Frequencies on Idaho's Snake River Plains: Ecological and Management Implications," paper presented at the Symposium on Cheatgrass Invasion, Shrub Die-Off, and Other Aspects of Shrub Biology and Management, Las Vegas, NV, April 5–7, 1989, https://citeseerx.ist.psu.edu/document?repid=rep1&type=pdf&doi=8541c17fc4bdee3331b67cfa94c3ab4d00d44191.

44. Personal conversation with Kumeyaay elder Michael Connelly Miskwish, El Campo, CA, 2022.

45. O. G. Davidson, "Climate Change Threatens an Iconic Desert Tree," *National Geographic*, October 28, 2015, https://www.nationalgeographic.com/science/article/151028-joshua-tree-climate-change-mojave-desert.

46. R. C. Averill-Murray, T. C. Esque, L. J. Allison, S. Bassett, S. K. Carter, K. E. Dutcher, S. J. Hromada, K. E. Nussear, and K. Shoemaker, *Connectivity of Mojave Desert Tortoise Populations: Management Implications for Maintaining a Viable Recovery Network*, 2021, US Geological Survey report prepared in cooperation with the US Fish and Wildlife Service, https://doi.org/10.3133/ofr20211033.

## 02.04. Abundance and Variation

1. H. A. Wright and A. W. Bailey, *Fire Ecology: United States and Southern Canada* (Hoboken, NJ: Wiley, 1982).

2. L. F. DeBano, D. G. Neary, and P. F. Ffolliott, *Fire's Effects on Ecosystems* (Hoboken, NJ: Wiley, 1998).

3. J. L. Lyon, H. S. Crawford, E. Czuhui, R. L. Fredricksen, R. F.

Harlow, L. J. Metz, and H. A. Pearson, *Effects of Fire on Fauna: A State-of-Knowledge Review*, Report GTR-WO-6 (Washington, DC: US Department of Agriculture, Forest Service, 1978).

4. "What Do Wild Animals Do in Wildfires?" *National Geographic*, September 12, 2020, https://www.nationalgeographic.com/environment/article/150914-animals-wildlife-wildfires-nation-california-science.

5. J. K. Smith, *Wildland Fire in Ecosystems: Effects of Fire on Fauna* (Ogden, UT: US Department of Agriculture, Forest Service, Rocky Mountain Research Station, 2000).

6. C. T. Hanson, R. L. Sherriff, R. L. Hutto, D. A. DellaSala, T. T. Veblen, and W. L. Baker, "Setting the Stage for Mixed and High-Severity Fire," in *The Ecological Importance of Mixed-Severity Fires: Nature's Phoenix*, ed. D. A. DellaSala and C. T. Hanson, 3–22 (Waltham, MA: Elsevier, 2015).

7. M. E. Swanson, N. M. Studevant, J. L. Campbell, and D. C. Donato, "Biological Associates of Early-Seral Pre-Forest in the Pacific Northwest," *Forest Ecology and Management* 324 (2014): 160–71, https://doi.org/10.1016/j.foreco.2014.03.046.

8. W. O. Wirtz, "Responses of Rodent Populations to Wildfire and Prescribed Fire in Southern California Chaparral," in *Brushfires in California Wildlands: Ecological and Resource Management*, ed. J. E. Keeley and T. Scott, 63–67 (Fairfield, WA: International Association of Wildland Fire, 1995).

9. C. E. Bock and J. H. Bock, "On the Geographical Ecology and Evolution of Three-Toed Woodpeckers, *Picoides tridactylus* and *P. arcticus*," *American Midland Naturalist* 92, no. 2 (1974): 397–405.

10. C. T. Hanson and T. Y. Chi, "Black-Backed Woodpecker Nest Density in the Sierra Nevada, California," *Diversity* 12, no. 10 (2020): 364, https://doi.org/10.3390/d12100364.

11. D. C. Odion, C. T. Hansen, A. Arsenault, W. L. Baker, D. A. DellaSala, R. L. Hutto, W. Klenner, M. A. Moritz, R. L. Sherriff, T. T. Veblen, and M. A. Williams, "Examining Historical and Current Mixed-Severity Fire Regimes in Ponderosa Pine and Mixed-Conifer

Forests of Western North America," *PLoS ONE* 9, no. 2 (2014): e87852, https://doi.org/10.1371/journal.pone.0087852.

12. M. L. Bond, D. E. Lee, R. B. Siegel, and J. P. Ward Jr., "Habitat Use and Selection by California Spotted Owls in a Postfire Landscape," *Journal of Wildlife Management* 73, no. 7 (2009): 1116–24, https://www.jstor.org/stable/20616768.

13. M. L. Bond, D. E. Lee, R. B. Siegel, and M. W. Tingley, "Diet and Home-Range Size of California Spotted Owls in a Burned Forest," *Western Birds* 44, no. 2 (2013): 114–26, https://westernfieldornithologists.org/publications/journal/journal-volume-44-2/v44-2-bond-spotted_owls.

## 02.05. Dependency and Enhancement

1. A. G. Merrill, A. E. Thode, A. M. Weill, J. A. Fites-Kaufman, A. F. Bradley, and T. J. Moody, "Fire and Plant Interactions," box 8.1, "Example of Biological and Ecological Levels of Organization for Bishop Pine," in *Fire in California's Ecosystems*, 2nd ed., ed. J. W. van Wagtendonk, N. G. Sugihara, S. L. Stephens, A. E. Thode, K. E. Shaffer, and J. A. Fites-Kaufman, 103–22 (Berkeley: University of California Press, 2018), 118.

2. P. H. Zedler, C. R. Gautier, and G. S. McMaster, "Vegetation Change in Response to Extreme Events: The Effect of Short Interval between Fires in California Chaparral and Coastal Scrub," *Ecology* 64, no. 4 (1983): 809–18, https://doi.org/10.2307/1937204.

3. T. R. Plumb and A. P. Gomez, *Five Southern California Oaks: Identification and Postfire Management*, General Technical Report PSW-71 (Berkeley, CA: USDA, Forest Service, Pacific Southwest Forest and Range Experiment Station, 1983), https://doi.org/10.2737/PSW-GTR-71.

4. Y. Valachovic, L. Quinn-Davidson, and R. B. Standiford, "Can the California Forest Practice Rules Adapt to Address Conifer Encroachment?" in *Proceedings of the 7th California Oak Symposium: Managing Oak Woodlands in a Dynamic World*, General Technical Report PSW-GTR-251, R. B. Standiford and K. L. Purcell, technical coordinators,

515–20 (Fresno, CA: USDA, Forest Service, Pacific Southwest Research Station, 2015), https://www.fs.usda.gov/psw/publications/documents/psw_gtr251/psw_gtr251.pdf.

5. D. M. Swezy and J. K. Agee, "Prescribed Fire Effects on Fine Root and Tree Mortality in Old Growth Ponderosa Pine," *Canadian Journal of Forest Research* 21, no. 5 (1991): 626–34, https://doi.org/10.1139/x91-086.

6. R. J. Vogl, "Fire Adaptations of Some Southern California Plants," *Proceedings Tall Timbers Fire Ecology Conference* 7 (1968): 79–109, https://talltimbers.org/wp-content/uploads/2018/09/79-Vogl1967_op-1.pdf.

7. P. R. Gagnon, H. A. Passmore, W. J. Platt, J. A. Myers, C.E.T. Paine, and K. E. Harms, "Does Pyrogenicity Protect Burning Plants?" *Ecology* 91, no. 12 (2010): 3481–86, https://doi.org/10.1890/10-0291.1.

8. W. W. Wagener, "Guidelines for Estimating the Survival of Fire-Damaged Trees in California," USDA Forest Service Miscellaneous Paper PSW-MP-60 (Berkeley, CA: Pacific Southwest Forest and Range Experiment Station, 1961).

9. K. W. Spalt and W. E. Reifsnyder, *Bark Characteristics and Fire Resistance: A Literature Survey*, USDA Forest Service Occasional Paper SO-OP-193 (New Orleans, LA: USDA, Forest Service, Southern Forest Experiment Station, in cooperation with School of Forestry, Yale University, 1962).

10. D. L. Fry, J. Dawson, and S. L. Stephens, "Age and Structure of Mature Knobcone Pines in the Northern California Coast Range," *Fire Ecology* 8 (2012): 49–62, https://doi.org/10.4996/fireecology.0801049.

11. P. J. Clarke, M. J. Lawes, J. J. Midgley, B. B. Lamont, F. Ojeda, G. E. Burrows, N. J. Enright, and J. E. Knox, "Resprouting as a Key Functional Trait: How Buds, Protection and Resources Drive Persistence after Fire," *New Phytologist* 197, no. 1 (2013): 19–35, https://doi.org/10.1111/nph.12001.

12. S. C. Barro and S. G. Conrad, "Fire Effects on California Chaparral System: An Overview," *Environment International* 17, nos. 2–3 (1991): 135–49, https://doi.org/10.1016/0160-4120(91)90096-9.

13. W. J. Bond and B. W. van Wilden, *Fire and Plants* (New York: Chapman and Hall, 1996).

14. C. Tyler and M. I. Borchert, "Reproduction and Growth of the Chaparral Geophyte, *Zigadenus fremontii* (Liliaceae), in Relation to Fire," *Plant Ecology* 165 (2002): 11–20, https://doi.org/10.1023/A:1021460025277.

## PART THREE: FIRE PRINCIPLES

### 03.01. World on Fire

1. A. C. Mulkern, "Today's Climate Change Proves Much Faster Than Changes in the Past 65 Million Years," *Scientific American*, August 2, 2013, https://www.scientificamerican.com/article/todays-climate-change-proves-much-faster-than-changes-in-past-65-million-years.

2. J. R. Marlon, P. J. Bartlein, M. K. Walsh, S. P. Harrison, K. J. Brown, M. E. Edwards, P. E. Higuera, M. J. Power, R. S. Anderson, C. Briles, A. Brunelle, C. Carcaillet, M. Daniels, F. S. Hu, M. Lavoie, C. Long, T. Minckley, P.J.H. Richard, A. C. Scott, D. S. Shafer, W. Tinner, C. E. Umbanhowar Jr., and C. Whitlock, "Wildfire Responses to Abrupt Climate Change in North America," *Proceedings of the National Academy of Sciences* 106, no. 8 (2009): 2519–24, https://doi.org/10.1073/pnas.0808212106.

3. M. A. Krawchuk, and M. A. Moritz, *Fire and Climate Change in California*, California Energy Commission, Publication no. CEC-500-2012-026, 2012, https://www.pepperwoodpreserve.org/wp-content/uploads/2016/03/Moritz-and-Krawchuk-2012-Changes-in-CA-fire-frequency-1911-2099-CEC-500-2012-026.pdf.

4. M. F. Seleiman, N. Al-Suhaibani, N. Ali, M. Akmal, M. Alotaibi, Y. Refay, T. Dindaroglu, H. H. Abdul-Wajid, and M. L. Battaglia, "Drought Stress Impacts on Plants and Different Approaches to Alleviate Its Adverse Effects," *Plants* 10, no. 2 (2021): 259, https://doi.org/10.3390/plants10020259.

5. J. Robbins, "Climate Whiplash: Wild Swings in Extreme Weather Are on the Rise," *Yale Environment* 360, November 14, 2019, https://e360.yale.edu/features/climate-whiplash-wild-swings-in-extreme-weather-are-on-the-rise.

6. D. R. Cayan, E. P. Maurer, M. D. Dettinger, M. Tyree, and K. Hayhoe, "Climate Change Scenarios for the California Region," *Climate Change* 87, no. 1 (2008): 521–42, https://doi.org/10.1007/s10584-007-9377-6.

7. A. Thompson, "Lightning May Increase with Global Warming," *Scientific American*, November 23, 2014, https://www.scientificamerican.com/article/lightning-may-increase-with-global-warming.

8. Cayan et al., "Climate Change Scenarios."

9. J. H. Thorne, R. M. Boynton, L. E. Flint, and A. L. Flint, "The Magnitude and Spatial Patterns of Historical and Future Hydrological Change in California's Watersheds," *Ecosphere 6*, no. 2 (2015): 1–30, https://doi.org/10.1890/ES14-00300.1.

10. A. L. Westerling, H. G. Hidalgo, D. R. Cayan, and T. W. Swetnam, "Warming and Earlier Spring Increases Western U.S. Forest Wildfire Activity," *Science* 313, no. 5789 (2006): 940–43, https://doi.org/10.1126/science.1128834.

11. A. Bhatia and N. Popovich, "These Maps Tell the Story of Two Americas: One Parched, One Soaked," *New York Times*, August 24, 2021, https://www.nytimes.com/interactive/2021/08/24/climate/warmer-wetter-world.html.

12. California Air Resources Board, "California Wildfire Emission Estimates," 2021, https://ww2.arb.ca.gov/sites/default/files/classic/cc/inventory/Wildfire%20Emission%20Estimates%202000-2021.pdf.

13. Deep Carbon Observatory, "Scientists Quantify Global Volcanic CO2 Venting; Estimate Total Carbon on Earth," Phys.org, October 1, 2019, https://phys.org/news/2019-10-scientists-quantify-global-volcanic-co2.html.

14. K. Kerlin, "Grasslands More Reliable Carbon Sink Than Trees," UC Davis, July 9, 2018, https://climatechange.ucdavis.edu/climate/news/grasslands-more-reliable-carbon-sink-than-trees.

15. R. Jordan, "Stanford-Led Study Reveals a Fifth of California's Sierra Nevada Conifer Forests Are Stranded in Habitats That Have Grown Too Warm for Them," Stanford University, February 28, 2023, https://news.stanford.edu/2023/02/28/zombie-forests.

16. A. P. Hill, C. J. Nolan, K. S. Hemes, T. W. Cambron, and C. B. Field, "Low-Elevation Conifers in California's Sierra Nevada Are out of Equilibrium with Climate," *PNAS Nexus* 2, no. 2 (2023): pgad004, https://doi.org/10.1093/pnasnexus/pgad004.

## 03.02. Fire Is the Hunter

1. E. Schlickman and B. Milligan, *Design by Fire: Resistance, Co-Creation and Retreat in the Pyrocene* (New York: Routledge, 2023).

2. CAL FIRE, "Statistics," accessed March 4, 2024, https://www.fire.ca.gov/our-impact/statistics.

3. L. S. Pile, M. D. Meyer, R. Rojas, O. Roe, and M. T. Smith, "Drought Impacts and Compounding Mortality on Forest Trees in the Southern Sierra Nevada," *Forests* 10, no. 3 (2019): 237, https://doi.org/10.3390/f10030237.

4. V. R. Kane, B. N. Bartl-Geller, G. R. Cova, C. P. Chamberlain, L. van Wagtendonk, and M. P. North, "Where Are the Large Trees? A Census of Sierra Nevada Large Trees to Determine Their Frequency and Spatial Distribution across Three Large Landscapes," *Forest Ecology and Management* 546 (2023): 121351, https://doi.org/10.1016/j.foreco.2023.121351.

5. Fire and Resource Assessment Program of the California Department of Forestry and Fire Protection, *California's Forests and Rangelands: 2017 Assessment*, CAL FIRE, 2018, https://cdnverify.frap.fire.ca.gov/media/4babn5pw/assessment2017.pdf.

## 03.03. A Policy of Prescription

1. N. S. Diffenbaugh and C. B. Field, "Changes in Ecologically Critical Terrestrial Climate Conditions," *Science* 341, no. 6145 (2013): 486–92, https://doi.org/10.1126/science.1237123.

2. H. H. Biswell, *Prescribed Burning in California Wildland Vegetation Management* (Berkeley: University of California Press, 1989).

3. R. K. Miller, C. B. Field, and K. J. Mach, "Barriers and Enablers for Prescribed Burns for Wildfire Management in California," *Nature Sustainability* 3 (2020): 101–109, https://doi.org/10.1038/s41893-019-0451-7.

4. S. Pyne, *Fire: A Brief History* (Seattle: University of Washington Press, 2001).

5. M. Kaufmann, A. Shlisky, and P. Marchand, "Good Fire, Bad Fire: How to Think about Forest Land Management and Ecological Processes," Conservation Gateway, October 25, 2010, https://www.conservationgateway.org/Files/Pages/good-fire-bad-fire-how-th.aspx.

6. Fire and Resource Assessment Program of the California Department of Forestry and Fire Protection, *California's Forests and Rangelands: 2017 Assessment*, CAL FIRE, 2018, https://cdnverify.frap.fire.ca.gov/media/4babn5pw/assessment2017.pdf.

7. CAL FIRE, "Fuels Reduction," accessed March 4, 2024, https://www.fire.ca.gov/what-we-do/natural-resource-management/fuels-reduction.

8. S. L. Stephens, S. J. Husari, T. Nichols, N. G. Sugihara, and B. M. Collins, "Fire and Fuel Management," in *Fire in California's Ecosystems*, ed. J. W. Van Wagtensdonk, N. G. Sugihara, S. L. Stephens, A. E. Thode, K. E. Shaffer, and J. A. Fites-Kaufmann, 411–28 (Berkeley: University of California Press, 2018).

9. M. North, J. Innes, and H. Zald, "Comparison of Thinning and Prescribed Fire Restoration Treatments to Sierran Mixed-Conifer Historic Conditions," *Canadian Journal of Forest Research* 37 (2007): 331–42, https://doi.org/10.1139/X06-236.

10. C. Swimmer, "*The Mendocino Trail Stewards: When Does a Struggle Become a Movement?*" Trees Foundation, July 2021, https://treesfoundation.org/2021/07/then-now-campaign-to-restore-jackson-state-forest/.

11. R. Sabalow and D. Kasler, "'Self-Serving Garbage': Wildfire Experts Escalate Fight over Saving California Forests," *Sacramento Bee*, October 25, 2021, reposted by the Sierra Nevada Research Institute, https://snri.ucmerced.edu/news/2021/self-serving-garbage-wildfire-experts-escalate-fight-over-saving-california-forests.

12. J. K. Agee and C. N. Skinner, "Basic Principles of Fuel Reduction Treatments," *Forest Ecology and Management* 211, nos. 1–2 (2005): 45–56, https://doi.org/10.1016/j.foreco.2005.01.034.

13. A. L. Westerling, B. P. Bryant, H. K. Preisler, T. P. Holmes, H. G. Hidalgo, T. Das, and S. R. Shresta, "Climate Change and Growth

Scenarios for California Wildfire," *Climatic Change* 109 (2011): 445–63, https://doi.org/10.1007/s10584-011-0329-9.

14. Little Hoover Commission, "Fire on the Mountain: Rethinking Forest Management in the Sierra Nevada," Report 242, February 2018, https://lhc.ca.gov/report/fire-mountain-rethinking-forest-management-sierra-nevada/.

15. B. M. Collins and G. B. Roller, "Early Forest Dynamics in Stand-Replacing Fire Patches in the Northern Sierra Nevada, California," *Landscape Ecology* 28 (2013): 1801–13, https://doi.org/10.1007/s10980-013-9923-8.

16. T. W. McGiniis, J. E. Keeley, S. L. Stephens, and G. B. Roller, "Fuel Buildup and Potential Fire Behavior after Stand-Replacing Fires, Logging Fire-Killed Trees and Herbicide Shrub Removal in Sierra Nevada Forests," *Forest Ecology and Management* 260, no. 1 (2010): 22–35, https://doi.org/10.1016/j.foreco.2010.03.026.

17. S. Pyne, "Pyric Other, Pyric Double: Fire Time, Fire Feral, Fire Extinct," *Australian Humanities Review* 52 (2012): 199–203, https://press-files.anu.edu.au/downloads/press/p196961/pdf/13-Pyne.pdf.

18. J. W. van Wagtendonk, "The History and Evolution of Wildland Fire Use," *Fire Ecology* 3 (2007): 3–17, https://doi.org/10.4996/fireecology.0302003.

## 03.04. Repair, Restore, and Reciprocate

1. California Senate Bill 337, 2023, https://legiscan.com/CA/text/SB337/2023.

2. Administration of Governor Gavin Newsom, *Pathways to 30x30 California: Accelerating Conservation of California's Nature*, 30x30 California, April 22, 2022, https://canature.maps.arcgis.com/sharing/rest/content/items/8da9faef231c4e31b651ae6dff95254e/data.

3. "GAP Analysis Project," USGS, n.d., https://www.usgs.gov/programs/gap-analysis-project.

4. GreenInfo Network, "Including GAP Codes in CPAD and CCED," n.d., https://www.calands.org/wp-content/uploads/2022/12/Method-for-Including-Gap-Codes-in-CPAD-and-CCED.pdf.

5. California Department of Fish and Wildlife, "Science: Habitat Connectivity," n.d., https://wildlife.ca.gov/Science-Institute/Habitat-Connectivity.

# Selected Bibliography


Agee, J. K. 1993. *Fire Ecology of Pacific Northwest Forests*. Washington, DC: Island Press.

Anderson, M. K. 2005. *Tending the Wild: Native American Knowledge and the Management of California's Natural Resources*. Berkeley: University of California Press.

Atwater, B. F., C. W. Hedel, and E. J. Helley. 1977. "Late Quaternary Depositional History, Holocene Sea-Level Changes, and Vertical Crustal Movement, Southern San Francisco Bay, California." US Geological Survey Professional Paper 1014. https://pubs.usgs.gov/pp/1014/p1014_text.pdf.

Bakker, E., and G. Slack. 1971. *An Island Called California: An Ecological Introduction to Its Natural Communities*. Berkeley: University of California Press.

Baldwin, B. G., D. H. Goldman, D. J. Keil, R. Patterson, T. J. Rosatti, and D. H. Wilken, eds. 2012. *The Jepson Manual: Vascular Plants of California*. 2nd ed. Berkeley: University of California Press. (The Jepson eFlora, the Jepson Herbarium, University of California, Berkeley, ucjeps.berkeley .edu/eflora/.)

Barbour, M. G., T. Keeler-Wolf, and A. A. Schoenherr. 2007. *Terrestrial Vegetation of California*. 3rd ed. Berkeley: University of California Press.

Barbour, M. G., and J. Major, eds. 1977. *Terrestrial Vegetation in California*. Hoboken, NJ: Wiley.

Barbour, M. G., B. Pavlik, F. Drysdale, and S. Lindstrom. 1993. *California's Changing Landscapes: Diversity and Conservation of California Vegetation*. Sacramento: California Native Plant Society.

Beck, W. A., and Y. D. Haase. 1974. *Historical Atlas of California*. Norman: University of Oklahoma Press.

Beesley, D. 2004. *Crow's Range: An Environmental History of the Sierra Nevada*. Reno: University of Nevada.

Behrensmeyer, A. K., J. D. Damuth, W. A. DiMichele, R. Potts, H.-D. Dues, and S. L. Wing. 1992. *Terrestrial Ecosystems through Time:*

*Evolutionary Paleoecology of Terrestrial Plants and Animals*. Chicago: University of Chicago Press.

Bettinger, R. L. 2015. *Orderly Anarchy: Sociopolitical Evolution in Aboriginal California*. Berkeley: University of California Press.

Blackwelder, E. 1931. "Pleistocene Glaciation in the Sierra Nevada and Basin Ranges." *Geologic Society of America Bulletin* 42, no. 4: 865–922.

Bolsinger, C. L. 1988. *The Hardwoods of California's Timberlands, Woodlands, and Savannas*. Resource Bulletin PNW-RB-148. (Portland, OR: USDA, US Forest Service, Pacific Northwest Research Station.

Bond, W. J., and B. W. van Wilden. 1996. *Fire and Plants*. New York: Chapman and Hall.

Bowcutt, F. 2015. *The Tanoak Tree: An Environmental History of a Pacific Coast Hardwood*. Seattle: University of Washington Press.

Briles, C. E., C. Whitlock, C. N. Skinner, and J. Mohr. 2011. "Holocene Forest Development and Maintenance on Different Substrates in the Klamath Mountains, Northern California." *Ecology* 92, no. 3: 590–601.

British Columbia Ministry of Forests and British Columbia Ministry of Environment, Land and Parks. 1995. *Biodiversity Guidebook*. Victoria, BC: BC Forest Practices Code.

Cal Alive! 2009. *Habitats Alive! An Ecological Guide to California's Diverse Habitats*. Exploring Biodiversity Teachers Resource Guide. El Cerrito: California Institute for Biodiversity, and Claremont, CA: Rancho Santa Ana Botanic Garden.

California Department of Fish and Game. 2005. *California Wildlife: Conservation Challenges, California's Wildlife Action Plan*. Edited by D. Bunn, A. Mummert, M. Hoshovsky, K. Gilardi, and S. Shanks. Davis: UC Davis Wildlife Health Center.

California Department of Fish and Wildlife. n.d. "Wildlife Habitats—California Wildlife Habitat Relationships System" (CWHR), www.wildlife.ca.gov/Data/CWHR/Wildlife-Habitats.

California Interagency Wildlife Task Group. 2005. *Habitat Classification Rules: California Wildlife Habitat Relationships System*. California Department of Fish and Game. nrm.dfg.ca.gov/FileHandler.ashx?DocumentID=65851&inline.

Carle, D. 2004. *Introduction to Water in California*. California Natural History Guides. Berkeley: University of California Press.

Carle, D. 2008. *Introduction to Fire in California*. California Natural History Guides. Berkeley: University of California Press.

Carle, D. 2010. *Introduction to Earth, Soil and Land in California*. California Natural History Guides. Berkeley: University of California Press.

Cheng, S., ed. 2004. *Forest Service Research Natural Areas in California*. Gen. Tech. Rep. PSW-GTR-188. Albany, CA: Pacific Southwest Research Station, US Department of Agriculture, US Forest Service.

Childs, C. 2001. *The Secret Knowledge of Water: Discovering the Essence of the American Desert*. New York: Back Bay Books.

Clark, D. H. 1995. "Extent, Timing, and Climatic Significance of Latest Pleistocene and Holocene Glaciation in the Sierra Nevada, California." PhD diss., University of Washington. https//doi.org/10.2172/527434.

Cobb, R., D. M. Rizzo, K. J. Hayden, M. Garbelotto, J.A.N. Filipe, C. A. Gilligan, W. W. Dillon, R. K. Meentemeyer, Y. S. Valachovic, E. Goheen, T. J. Swiecki, E. M. Hansen, and S. J. Frankel. 2013. "Biodiversity Conservation in the Face of Dramatic Forest Disease: An Integrated Conservation Strategy for Tanoak (*Notholithocarpus densiflorus*) Threatened by Sudden Oak Death." *Madroño* 60, no. 2 (April 2013): 151–64. http://dx.doi.org/10.3120/0024-9637-60.2.151.

Crampton, B. 1974. *Grasses in California*. Berkeley: University of California Press.

Cunningham, L. 2010. *A State of Change: Forgotten Landscapes of California*. Berkeley, CA: Heyday.

DellaSala, D. E., and C. T. Hanson, eds. 2015. *The Ecological Importance of Mixed-Severity Fires: Nature's Phoenix*. Waltham, MA: Elsevier.

Didion, J. 2003. *Where I Was From*. New York: Random House.

Dirke-Edmunds, J. 1999. *Not Just Trees: The Legacy of a Douglas-Fir Forest*. Pullman: Washington State University Press.

DiTomaso, J. M., G. B. Kyser, S. R. Oneto, R. G. Wilson, S. B. Orloff, L. W. Anderson, S. D. Wright, J. A. Roncoroni, T. L. Miller, T. S. Prather, C. Ransom, K. G. Beck, C. Duncan, K. A. Wilson, and J. J. Mann. 2013. *Weed Control in Natural Areas in the Western United States*. UC Davis Weed Research and Information Center. Davis: University of California.

Dobzhansky, T. 1973. "Nothing in Biology Makes Sense Except in the

Light of Evolution," *American Biology Teacher*. Reprinted in J. P. Zetterberg, ed. 1983. *Evolution versus Creationism*. Phoenix, AZ: ORYX.

Dolanc, C. R., H. D. Safford, S. Z. Dobrowksi, and J. H. Thorne. 2014. "Twentieth-Century Shifts in Abundance and Composition of Vegetation Types in the Sierra Nevada, CA." *Applied Vegetation Science* 17: 442–55.

Durrenberger, R. W., and R. B. Johnson. 1976. *California: Patterns on the Land*. California Council for Geographic Education. Palo Alto, CA: Mayfield.

Egan, T. 2009. *The Big Burn: Teddy Roosevelt and the Fire That Saved America*. Boston: Houghton Mifflin Harcourt.

Faber, P. M., ed. 1997. *California's Wild Gardens: A Guide to Favorite Botanical Sites*. Berkeley: University of California Press.

Faith, J. T., and T. A. Surovell. 2009. "Synchronous Extinction of North America's Pleistocene Mammals." *PNAS* 106, no. 49: 20641–45. https://doi.org/10.1073/pnas.0908153106.

Farjon, A. 1998. *World Checklist and Bibliography of Conifers*. Kew, UK: Royal Botanical Gardens.

Farmer, J. 2017. *Trees in Paradise: The Botanical Conquest of California*. Berkeley, CA: Heyday.

Finson, B., ed. 1983. *Discovering California*. San Francisco: California Academy of Sciences.

Fites, J. 1993. "Ecological Guide to Mixed Conifer Plant Associations of the Northern Sierra Nevada and Southern Cascades." In *USDA Forest Service Technical Paper R5-ECOL-TP-*001. Albany, CA: USDA Forest Service, Pacific Southwest Region.

Fleischner, T. L., ed. 2011. *The Way of Natural History*. San Antonio, TX: Trinity University Press.

Forest Climate Action Team. 2018. *California Forest Carbon Plan: Managing Our Forest Landscapes in a Changing Climate*. Sacramento: California Natural Resources Agency.

Fradkin, P. L. 1995. *The Seven States of California: A Natural and Human History*. Berkeley: University of California Press.

Gonzales, A. G., and J. Hoshi, eds. 2015. *California State Wildlife Action Plan, 2015 Update: A Conservation Legacy for Californians*. Sacramento: California Department of Fish and Wildlife. wildlife.ca.gov/SWAP/Final.

Graeber, D., and D. Wengrow. 2021. *The Dawn of Everything: A New History of Humanity.* New York: Farrar, Straus and Giroux.

Griffin, J. R., and W. B. Critchfield. 1976. *The Distribution of Forest Trees in California.* USDA Forest Service Research Paper PSW-82. Albany, CA: USDA Forest Service, Pacific Southwest Forest and Range Experiment Station.

Grillos, S. J. 1966. *Ferns and Fern Allies of California.* Berkeley: University of California Press.

Gudde, E. G. 1949. *California Place Names: The Origin and Etymology of Current Geographical Names.* Berkeley: University of California Press.

Hanson, M. 2014. *A Species Guide to the Berryessa Snow Mountain Region.* Woodland, CA: Tuleyome.

Harden, D. 1992. *California Geology.* New York: Pearson.

Hart, J. 1975. *Hiking the Bigfoot Country: The Wildlands of Northern California and Southern Oregon.* San Francisco: Sierra Club.

Hart, J. 1978. *San Francisco's Wilderness Next Door.* San Rafael, CA: Presidio.

Hart, J. D. 1987. *A Companion to California.* Berkeley: University of California.

Henson, P., and D. J. Usner. 1993. *The Natural History of Big Sur.* Berkeley: University of California Press.

Hickman, J. C. 1993. *The Jepson Manual: Higher Plants of California.* Berkeley: University of California Press.

Hilty, J. A., A.T.H. Keeley, W. Z. Lidicker, and A. M. Merenlender. 2019. *Corridor Ecology: Linking Landscapes for Biodiversity Conservation and Climate Adaptation.* 2nd ed. Washington, DC: Island Press.

Holing, D. 1988. *California Wild Lands: A Guide to the Nature Conservancy Preserves.* San Francisco: Chronicle Books.

Holland, R. F. 1986. *Preliminary Descriptions of the Terrestrial Natural Communities of California.* Sacramento: California Department of Fish and Game.

Holland, V. L., and D. J. Keil. 1987; 1995. *California Vegetation.* San Luis Obispo: El Corral, California Polytechnic State University; Dubuque, IA: Kendall Hunt (1995).

Hornbeck, D., and P. S. Kane. 1983. *California Patterns: A Geographical and Historical Atlas.* Palo Alto, CA: Mayfield.

Huntsinger, L. 2002. *Sierra Nevada Grazing in Transition: The Role of Forest Service Grazing in the Foothill Ranches of California*. Sierra Nevada Alliance.

Jenks, M. A. 2011. *Plant Nomenclature*. West Lafayette, IN: Department of Horticulture and Landscape Architecture, Purdue University.

Jensen, H. A. 1947. *A System for Classifying Vegetation in California*. Sacramento: California Department of Fish and Game.

Johnson, P. 1970. *Pictorial History of California*. New York: Bonanza Books.

Johnston, V. R. 1998. *Sierra Nevada: The Naturalist's Companion*. Berkeley: University of California Press.

Jones, P. D., K. R. Briffa, T. P. Barnett, and S.F.B. Tett. 1998. "High-Resolution Paleoclimatic Records for the Last Millennium: Interpretation, Integration and Comparison with General Circulation Model Control-Run Temperatures." *Holocene* 8: 455–71.

Karuk Tribe Department of Natural Resources. 2019. "Karuk Climate Adaptation Plan." karuktribeclimatechangeprojects.files.wordpress .com/2019/10/reduced-size_final-karuk-climate-adaptation-plan.pdf.

Kauffman, E. 2003. *Atlas of Biodiversity of California*. Sacramento: State of California, the Resources Agency, Department of Fish and Game.

Kauffmann, M. E. 2013. *Conifers of the Pacific Slope: A Field Guide to the Conifers of California, Oregon, and Washington*. Kneeland, CA: Backcountry Press.

Keator, G. 2009. *California Plant Families West of the Sierran Crest and Deserts*. Berkeley: University of California Press.

Kimmerer, R. W. 2013. *Braiding Sweetgrass: Indigenous Wisdom, Scientific Knowledge, and the Teachings of Plants*. Minneapolis: Milkweed Editions.

King, G. 2023. *The Ghost Forest: Racists, Radicals, and Real Estate in the California Redwoods*. New York: Public Affairs.

Krell, D., ed. 1979. *The California Missions: A Pictorial History*. Menlo Park, CA: Sunset Publishing.

Lanner, R. 1999. *Conifers of California*. Los Olivos, CA: Cachuma Press.

Lanner, R. M., and T. R. van Devender. 1998. "The Recent History of Pinyon Pines in the American Southwest." In *Ecology and Biography of Pinus*, edited by D. M. Richardson. Cambridge: Cambridge University Press.

Laws, J. M. 2007. *The Laws Field Guide to the Sierra Nevada*. California Academy of Sciences. Berkeley, CA: Heyday.

Lentz, J. E. 2013. *A Naturalist's Guide to the Santa Barbara Region*. Berkeley, CA: Heyday.

Leopold, A. S. 1985. *Wild California: Vanishing Lands, Vanishing Wildlife*. Berkeley: University of California Press.

Lightfoot, K. G. 2005. *Indians, Missionaries, and Merchants: The Legacy of Colonial Encounters on the California Frontiers*. Berkeley: University of California Press.

Lopez, B., and D. Gwartney. 2006. *Home Ground: A Guide to the American Landscape*. San Antonio, TX: Trinity University Press.

Lovelock, J. 1979. *Gaia*. Oxford: Oxford University Press.

Lyons, K., and M. B. Cuneo-Lazaneo. 1988. *Plants of the Coast Redwood Region*. Soquel, CA: Shoreline.

MacArthur, R. H., and E. O. Wilson. 1967. *The Theory of Island Biogeography*. Princeton, NJ: Princeton University Press.

Madley, B. 2017. *An American Genocide: The United States and the California Indian Catastrophe*. New Haven, CT: Yale University Press.

Margolin, M. 1974. *The East Bay Out: A Personal Guide to the East Bay Regional Parks*. Berkeley, CA: Heyday.

Margolin, M. 1978. *The Ohlone Way: Indian Life in the San Francisco–Monterey Bay Area*. Berkeley, CA: Heyday.

Marianchild, K. 2014. *Secrets of the Oak Woodlands: Plants and Animals among California's Oaks*. Berkeley: Heyday.

Maul, D. C. 1958. *Silvical Characteristics of White Fir*. Albany, CA: Pacific Southwest Research Station, USDA, US Forest Service.

Mayer, K. E., and W. F. Laudenslayer Jr., eds. 1988. *A Guide to Wildlife Habitats in California*. Sacramento: California Department of Forestry and Fire Protection. (Updated online as California Department of Fish and Wildlife, "Wildlife Habitats—California Wildlife Habitat Relationships System," www.wildlife.ca.gov/Data/CWHR/Wildlife-Habitats.)

McPhee, J. 1993. *Assembling California*. New York: Farrar, Straus and Giroux.

Miller, C. S., and R. S. Hyslop. 1983. *California: The Geography of Diversity*. Mountain View, CA: Mayfield.

Mooney, H., E. Zavaleta, and M. C. Chapin, eds. 2016. *Ecosystems of California*. Oakland: University of California Press. www.jstor.org/stable/10.1525/j.ctv1xxzp6.

Moore, G., B. Kershner, C. Tufts, and D. Mathews. 2008. *National Wildlife Federation Field Guide to Trees of North America*. New York: Sterling.

Muir, J. 1876; 1984. Robert Engberg, ed. *Summering in the Sierra*. Madison: University of Wisconsin Press.

Muir, J. 1894; 1988. *The Mountains of California*. San Francisco: Sierra Club.

Munz, P. A., and D. D. Keck. 1959. *A California Flora*. Berkeley: University of California Press.

Nixon, S. 1966. *Redwood Empire*. New York: Galahad.

Noy, G., and R. Heide, eds. 2010. *The Illuminated Landscape: A Sierra Nevada Anthology*. Rocklin, CA: Sierra College Press, and Berkeley, CA: Heyday.

Ornduff, R. 1974. *An Introduction to California Plant Life*. Berkeley: University of California Press.

Ornduff, R., P. M. Faber, and T. Keeler-Wolf. 2003. *Introduction to California Plant Life*. California Natural History Guides. Berkeley: University of California Press.

Pavlik, B. M. 2008. *The California Deserts: An Ecological Rediscovery*. Berkeley: University of California Press.

Pavlik, B. M., P. C. Muick, S. G. Johnson, and M. Popper. 1991. *Oaks of California*. Los Olivos, CA: Cachuma Press.

Perry, C. 1999. *Pacific Arcadia: Images of California 1600–1915*. New York: Oxford University Press.

Peterson, B. 1993. *California: Vanishing Habitats and Wildlife*. Wilsonville, OR: Beautiful America.

Petrides, G. A., and O. Petrides. 1998. *Western Trees*. Peterson Field Guide. New York: Houghton Mifflin.

Plumb, T. R., and P. M. McDonald. 1981. *Oak Management in California*. General Technical Report PSW-54. Albany, CA: US Department of Agriculture, Forest Service, Pacific Southwest Forest and Range Experimental Station.

Polakovic, G. 1999. "Channel Island Woman's Bones May Rewrite History." *Los Angeles Times*, April 11. www.latimes.com/archives/la-xpm-1999-apr-11-mn-26401-story.html.

Potter, D., M. Smith, T. Beck, B. Kermeen, W. Hance, and S. Robertson. 1992. "Ecological Characteristics of Old Growth Lodgepole Pine in California." Unpublished report. Albany, CA: US Department of Agriculture, Forest Service, Pacific Southwest Region.

Pratt-Bergstrom, B. 2016. *When Mountain Lions Are Neighbors: People and Wildlife Working It Out in California*. Berkeley, CA: Heyday.

Press, D. 2002. *Saving Open Space: The Politics of Local Preservation in California*. Berkeley: University of California Press.

Pyne, S. 2001. *Fire: A Brief History*. Seattle: University of Washington Press.

Pyne, S. 2021. *The Pyrocene*. Berkeley: University of California Press.

Quinn, R. D., and S. C. Keeley. 2006. *Introduction to California Chaparral*. California Natural History Guides. Berkeley: University of California Press.

Reisner, M. 1986. *Cadillac Desert: The American West and Its Disappearing Water*. New York: Penguin.

Ritter, M. 2011. *A Californian's Guide to the Trees among Us*. Berkeley, CA: Heyday.

Ritter, M. 2018. *California Plants: A Guide to Our Iconic Flora*. San Luis Obispo, CA: Pacific Street Publishing.

Sarris, G. 2017. *How a Mountain Is Made*. Berkeley, CA: Heyday.

Sawyer, J. O., T. Keeler-Wolf, and J. Evens. 2009. *A Manual of California Vegetation*. 2nd ed. Sacramento: California Native Plant Society. https://www.cnps.org/vegetation/manual-of-california-vegetation; vegetation.cnps.org; www.wildlife.ca.gov/Data/VegCAMP/Natural-Communities.

Schoenherr, A. A. 1992. *A Natural History of California*. Berkeley: University of California Press.

Schonewald-Cox, C. M., S. M. Chambers, B. MacBryde, and L. Thomas, eds. 1983. *Genetics and Conservation*. Menlo Park, CA: Benjamin/Cummings.

Snyder, G. 1995. *A Place in Space: Ethics, Aesthetics and Watersheds*. Berkeley, CA: Counterpoint.

Snyder, G. 1996. *Mountains and Rivers without End*. Berkeley, CA: Counterpoint.

Spencer, W. D., P. Beier, K. Penrod, K. Winters, C. Paulman, H. Rustigan-Romsos, J. Strittholt, M. Parisi, and A. Pettler. 2010. *California Essential Habitat Connectivity Project: A Strategy for Conserving a Connected California*. Prepared for the California Department of Transportation, California Department of Fish and Game, and Federal Highways Administration. wildlife.ca.gov/Conservation/Planning/Connectivity/CEHC.

Starr, K. 2005. *California: A History*. New York: Modern Library.

Stegner, W. 1946. *The Sound of Mountain Water*. New York: Dutton.

Stegner, W. 1971. *Angle of Repose*. New York: Penguin.

Storer, T. I., and R. L. Usinger. 2004. *Sierra Nevada Natural History*. Rev. ed. California Natural History Guide Series No. 73. Berkeley: University of California Press.

Timbrook, J. 2007. *Chumash Ethnobotany: Plant Knowledge among the Chumash People of Southern California*. Santa Barbara, CA: Santa Barbara Museum of Natural History, and Berkeley, CA: Heyday.

Tweed, W. C. 2016. *King Sequoia: The Tree That Inspired a Nation, Created Our National Park System, and Changed the Way We Think about Nature*. Berkeley, CA: Heyday.

USDA Forest Service. 2013. *Ecological Restoration Implementation Plan*. R5-MB-249. Vallejo, CA: USDA Forest Service, Pacific Southwest Region. https://www.fs.usda.gov/Internet/FSE`DOCUMENTS/stelprdb5411383.pdf.

Vaillant, J. 2023. *Fire Weather*. New York: Knopf.

van Wagtendonk, J. W., S. L. Sugihara, S. L. Stephens, A. E. Thode, K. E. Shaffer, and J. A. Fites-Kaufman, eds. 2018. *Fire in California's Ecosystems*. 2nd ed. Berkeley: University of California Press.

Wallace, D. R., 1983. *The Klamath Knot: Explorations of Myth and Evolution*. Berkeley: University of California Press.

Wallace, D. R. 2014. *Articulate Earth: Adventures in Ecocriticism*. Kneeland, CA: Backcountry Press.

Wallace, D. R. 2015. *Mountains and Marshes: Exploring the Bay Area's Natural History*. Berkeley, CA: Counterpoint.

White, M. 1983. *Trinity Alps and Vicinity*. Berkeley, CA: Wilderness Press.

Whittaker, R. H. 1975. *Communities and Ecosystems*. 2nd rev. ed. New York: Macmillan.

Wilson, E. O. 2002. *The Future of Life*. New York: Knopf.

Self portrait 2024

# About the Author

Obi Kaufmann is the author of several award-winning, best-selling books on California's ecology, biodiversity, and geography. His first book, *The California Field Atlas*, recontextualized popular ideas about what he calls "California's more-than-human world." His next books—*The State of Water: Understanding California's Most Precious Resource* and the California Lands Trilogy (*The Forests of California*, *The Coasts of California*, and *The Deserts of California*)—present a comprehensive survey of California's physiography, biogeography, evolutionary past, and unfolding future. *The Deserts of California* won the 2024 Golden Poppy Award for California Lifestyle from the California Independent Booksellers Alliance.

Obi regularly travels around the state, presenting his work at venues from the Klamath-Siskiyou Wildlands Center to the Mojave Desert Land Trust, and he was the 2023 artist-in-residence for the National Park Service at Whiskeytown National Recreation Area. A lifelong resident of California, Obi Kaufmann makes his home base in Oakland when he isn't backpacking, and he is working on more field atlases to come. You can catch him every month in conversation with author and tribal chairman Greg Sarris on their podcast, *Place and Purpose*, and you can follow his work on Instagram at @coyotethunder.